Other volumes in this series:

VOLTAGE QUALITY
IN ELECTRICAL
POWER SYSTEMS

J. Schlabbach,
D. Blume and
T. Stephanblome

Translated by
M. Daly and P. Anderson,
Accura Translations

The Institution of Electrical Engineers

Originally published in German by VDE-Verlag
'Spannungsqualität in elektrischen Netzen'
© 1999: VDE-Verlag

This edition published by: The Institution of Electrical Engineers,
London, United Kingdom

English edition © 2001: The Institution of Electrical Engineers

The Institution of Electrical Engineers,
Michael Faraday House,
Six Hills Way, Stevenage,
Herts. SG1 2AY, United Kingdom

British Library Cataloguing in Publication Data
 Schlabbach, J.
 Voltage quality in electrical power systems.
 (IEE power series; no. 36)
 1. Voltage regulators 2. Electrical power systems – Quality control
 3. Electrical power system stability
 I. Title II. Blume, D. III. Stephanblome, T. IV. Institution of
 Electrical Engineers
 621.3′1

ISBN 0 85296 975 9

Typeset by RefineCatch Ltd, Bungay, Suffolk
Printed in England by MPG Books Ltd, Bodmin, Cornwall

Contents

Foreword

The problem of voltage quality is gaining increasing importance due to the widespread use of power electronics (increasing emitted interference) on the one hand, and the reduction in the signal levels in electronic equipment (increased interference susceptibility) on the other hand. The voltage quality depends on various phenomena of the network perturbations, one of which, the conducted disturbance, is examined in this book.

This book deals with the subject of voltage quality from a practical viewpoint but without omitting the mathematical aspects. The problems set out in this book are taken from many examples of operating practice and methods of solving these problems are indicated. Practical tasks and examples of applications given in individual chapters deal with the topic in more detail.

As this book is a translation from a German book, published in the VDE-Verlag, VDE classification is always mentioned. However the standards in this book are designated to the EN numbers and so far as EN numbers are not available to documents of the IEC. The original German book presents a separate chapter with a detailed reference list of the standards and VDE-specifications. As a detailed overview of the standards in relation to the VDE-specification will be of less use for the reader in the international market, the reader is referred to section 1.3 of this book, which indicates the general structure of standards, and to the publications of the International Electro-technical Commission IEC. The particular features of standardisation are dealt with in individual chapters where this appears practical. Status of the standards is based on the best knowledge of the authors in May 2001.

Chapter 1 is a general introduction to the subject. The mathematical basics required, in conjunction with the subject of voltage quality, are also refreshed. Chapter 2 outlines in detail the occurrence, calculation and effects of harmonics and intermediate harmonics in networks. Chapter 3 deals with voltage fluctuations and flicker. Flicker is calculated using empirical formulae for periodic changes and also for random signals. Chapter 4 covers causes, description and effects of asymmetries. The measurement and assessment of system perturbations are given in detail in Chapter 5, with general selection criteria for methods and systems of measurement being worked out. Chapter 6 gives measures for the improvement of voltage quality. This chapter deals not only with standard methods, but also shows how innovative methods and equipment may be used. Chapter 7 provides all the instructions for practical procedures, all of which are based on the comprehensive practical experience of the authors.

Within the context of this book, the three phases of the power system are named R, Y and B instead of L1, L2 and L3. The three components of the symmetrical components are named positive, negative and zero sequence system instead of positive, negative and zero phase-system for easier reading.

The book is addressed to engineers practising in industry, electric utilities and engineering companies in which questions of system perturbations arise. Parts of this book are suitable as an accompaniment to study documents for teachers and students at universities.

The authors wish to express their thanks to all the companies in whose networks and systems measurements were performed, the results of which are included in the application examples. We wish to thank Dipl.-Ing. Roland Werner from the VDE-VERLAG for his co-operation and assistance and Dr. Robin Mellors-Bourne from the IEE for his efforts to publish this book in the United Kingdom. Special thanks to Ms. Diana Levy from the IEE for spending a lot of time and effort to improve figures, captions and wording.

Jürgen Schlabbach,
University of Applied Sciences of Bielefeld;

Dirk Blume,
team GmbH, Herten;

Thomas Stephanblome,
EUS GmbH, Gelsenkirchen

September 2001

Chapter 1

Introduction

1.1 Electromagnetic compatibility in electrical supply systems

According to the 'Order on the general conditions for electrical supply to tariff customers (AVBEltV)', the 'Technical connecting conditions for connecting to the network (TAB)' and the contracts for special contract customers:

'Systems and consumer equipment are to be operated so that interference for other customers and perturbations on devices of the power supply company or third parties are precluded.'

This statement is supplemented by the definition of the terms of electromagnetic compatibility (EMC) according to VDE 0870 as:

'The ability of an electrical installation (equipment, device or system) to operate satisfactorily in its electromagnetic environment without introducing impermissible electromagnetic interference with respect to any part of this environment to which other installations also belong.'

Recognition of the problem of electromagnetic compatibility is not new. As far back as 1892 a law was passed in the German Reich, which can be regarded as the first EMC law [1], as follows:

'Electrical systems shall, if a disturbance in the operation of one line by another has occurred, or may occur, at the expense of that part which due to the later system or a subsequent change to its existing system causes this disturbance, or the danger of same, where possible be designed so that they do not have a disturbing effect.'

It is a widely-held view that the defined application of the procedures given in the standards is sufficient to achieve EMC, i.e. that this leads to a secure operation of electrotechnical systems with regard to electromagnetic interference. This is correct only up to a point, because the standardisation only stipulates that the requirements be met for standard cases. However, technical systems are so complex and diverse with regard to their design and operation that the specifications in the standard do not go deep enough and therefore have to be interpreted. The electrical energy and, particularly the voltage, has many changing features at the changeover point to the customer, which sometimes have a considerable disturbing effect on the possible utilisation. System perturbations are a main area of these voltage disturbances. They occur when equipment with a non-linear current-voltage curve or with an operating behaviour which is not steady state is operated in a system with a finite short-circuit power, i.e. in a system with a finite impedance.

The problem of system perturbations is becoming increasingly important due to the increased use of power electronics (increased emitted interference) on one hand and the reduction in the signal level in electronic equipment (increased disturbance sensitivity) on the other. Some typical values of signal levels of equipment used in measurement and control are worth mentioning.

Electromagnetic equipment	10^{-1}	to	10^{1} W
Analogue electronic equipment	10^{-3}	to	10^{-1} W
Digital equipment	10^{-5}	to	10^{-3} W

When considering system perturbations it should be assumed that the interests of the consumer and power supply operator must be harmonised. This means that economic aspects have to be considered alongside the technical boundary condition of the equipment and the needs of the consumer.

It is therefore generally not possible to increase the short-circuit power of the system regardless of other factors to reduce the impedance in order to minimise system perturbations. Economic and technical boundaries are determinant in this case. On the other hand, the equipment operated in the system cannot be provided with any desired level of interference immunity because the costs for this would increase considerably in line with the level of interference immunity.

Between these boundary conditions, a compromise must be found, on which all consumers can rely. This should hold good, particularly for future system changes, and enable proper functioning of the equipment and systems into the future.

The individual phenomena of system perturbations will be investigated separately, showing that questions of measurability, the analysis of possible effects on equipment, and the specification of suitable remedies, can lead to different solutions in each case.

Future development requires constant monitoring of system perturbations. The reasons for this are as follows.

- System changes and restructuring (e.g. increasing the amount of cabling in systems) as well as the increased use of systems for reactive power compensation lead to lower resonant frequencies of the system.
- Changes to the make-up of consumers (replacement of ohmic consumers by electronic equipment, e.g. in industrial heat technology) and changes in consumer behaviour (increased use of small electronic devices) lead to higher disturbance levels.

- Measures to reduce consumption by replacing conventional lighting equipment by compact fluorescent lamps increases disturbance levels.
- The use of unconventional current and voltage transformers (optical waveguides) leads to an improvement in the measurement of system perturbations.
- Development of new compensation and remedial measures enables cost-effective solutions for improvement in the voltage quality.

Because of the periodic and/or random deviations and the random behaviour of the disturbances in the electrical supply system, as shown in Figure 1.1, the level of a disturbance can only be given as a frequency distribution. Also, the interference immunity of individual devices is statistically distributed. Functional impairment or failures of equipment need to be considered only in the overlap area.

Emitted interference levels, e.g. EN 61000 3–2 (VDE 0838 part 2) and interference immunity test levels, e.g. EN 50178 (VDE 0160) are specified on the basis of this probability. A guide value is the compatibility level, e.g. according to EN 61000 2–2; IEC 1000–2–2 (VDE 0839 part 2–2) for public low voltage systems, i.e. a specified level in the system for which a specific probability of electromagnetic compatibility exists, as shown in Figure 1.2. From this it can be deduced that disturbance levels which exceed the compatibility level occur with a certain probability. The timing and magnitude of this probability will vary for different disturbance phenomena.

The phenomenon of the interference quantity can be influenced both by the generating end (disturbance emittance) and the disturbed end (disturbance sensitivity) as well as by changing the disturbance transmission. An essential condition for the analysis of disturbance phenomena and possible remedies is

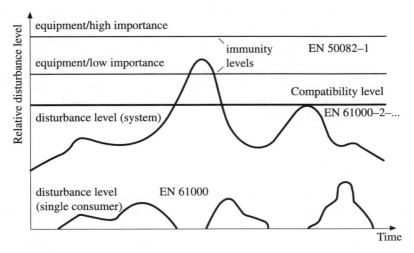

Figure 1.1   Random time course of disturbance, e.g. 5th harmonic
References to EN-norms to be understood as examples only

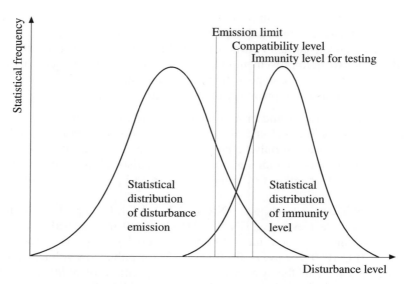

Figure 1.2   Probability distribution of disturbance level and immunity level

knowledge of the transmission mechanism between the disturbance source and disturbance drain. The basic relationships of the coupling mechanisms are summarised in Figure 1.3.

In this case, inductive, capacitive and galvanic couplings should be considered, provided the propagation times of the disturbances within the system under consideration can be ignored, e.g. the wavelength of the disturbance is large compared with the system dimensions. This quasi-steady state treatment applies to system perturbations in the field of electrical power supply systems.

On the other hand, couplings described by models of wave propagation or radiation influence should be considered if the wavelength of the disturbance is less than the system dimension, as is sometimes the case where there are EMC problems with electronic circuit boards. The same applies where the rise times of the disturbances are in the magnitude of the signal transit times. This is the case when considering pulse envelopes.

1.2 Classification of system perturbations

System perturbations occur as harmonic voltages, voltages with interharmonic frequencies, flicker, voltage changes, voltage change courses, voltage fluctuations and voltage asymmetries [2]. The determining frequency range for considering system perturbations extends in this case from zero-frequency quantities ($f = 0$ Hz) up to frequencies of $f \approx 10$ kHz. When considering the individual types of system perturbation, the particular frequency range is more closely considered. The individual phenomena of system perturbations are defined as follows.

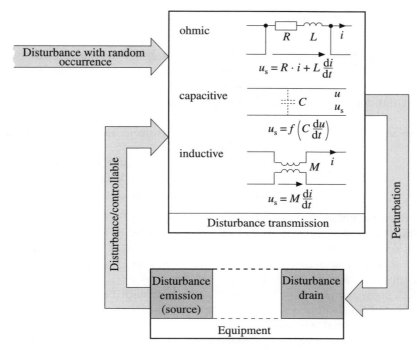

Figure 1.3   Basic relation of emission, transmission and coupling of disturbances

Harmonic	Sinusoidal oscillation whose frequency is a whole-number multiple of the fundamental frequency.
Interharmonic	Sinusoidal oscillation whose frequency is not a whole-number multiple of the fundamental frequency.
Flicker	Subjective impression of luminance fluctuation of filament lamps or fluorescent lamps.
Voltage change	Change to the r.m.s. value of the voltage.
Voltage change course	Time function of the difference between the r.m.s. value of the voltage at the start of the voltage change and the succeeding r.m.s. values.
Voltage fluctuation	Sequence of voltage changes or voltage change courses.
Voltage asymmetry	Deviation of three voltages of the three-phase system with regard to amplitude or deviation from the 120° phase difference.

Frequency fluctuations, i.e. deviations of the defined supply system nominal frequency are, on the contrary, a global phenomenon provided quasi-steady states are being considered.

In deviation from, or in addition to, the phenomena mentioned in the introduction, the following disturbance phenomena are defined in IEC 1000–2–1.

Short supply interruption	Interruptions in supply voltage for a maximum of one minute. Interpreted as a 'voltage sag' with 100% amplitude.
D.C. component	D.C. component of the voltage—at present under discussion in the IEC.
Mains signalling	Higher frequency transmission signals on high voltage lines. Audio-frequency telecontrols up to 2 kHz. PLC transmission up to 20 kHz. Telephone systems up 500 kHz.

It should be noted that voltage failures (short supply interruptions) do not represent system perturbations, but should be regarded as a disturbed operating state. Voltage failures are therefore not dealt with in this book. This does not mean that measures to reduce system perturbations cannot also be used as measures to deal with voltage failures.

D.C. components of the voltage are not dealt with because at present no stipulations for dealing with d.c. voltage components are given in either international or national standards.

The use of higher frequency signals on power supply lines for the purpose of signal transmission is dealt with in conjunction with the topic of harmonics and interharmonics. In this case, the field of audio-frequency telecontrol systems is of interest with regard to system perturbations.

1.3 EU directives, VDE specifications and standards

The topic of supply perturbations is at present dealt with in an extensive standards and regulations catalogue and represents a subdomain of electromagnetic compatibility. It is therefore incorporated in national legislation such as the EMC Act of 19.11.1992, the first EMC Amendment Act of 30.11.1995 and the Federal Clean Air Act of June 1996 by means of 26 implementation orders to date.

Electromagnetic compatibility is dealt with in the European context by EU directives, of which two should be mentioned here. EU directive 85–374 of July 1995 described electrical energy as a product for which a product liability is provided, from which the requirements to stipulate quality characteristics is derived. EU directive 89/336 of May 1989 described specifications for the EMC emissions from electrical systems.

The standardisation work in the field of electromagnetic compatibility is based on these directives, acts and orders. The previous voluntary application of VDE standards has grown in importance with harmonisation with EMC legislation and the corresponding European standards. Therefore, since the beginning of 1996 the CE mark, as an external indication of EMC conformity of equipment, must be carried on all equipment for sale in the EU. Manufacturers can perform their own test to enable the CE mark to be used. If there are no adequate test facilities available at the manufacturer, or if there are no relevant standards available, the CE mark can be awarded by an authorised certification organisation.

Standards are prepared by various specialist committees of the IEC, CENELEC and the CISPR. TC77 of the IEC is developing standards for measuring and test methods and for interference immunity for the complete frequency range, with the standardisation of the limits of emitted interference for the 0 Hz to 9 kHz frequency range being prepared by subcommittee SC77A and for the frequency range above 9 kHz by the CISPR. TC110 of CENELEC will then draw up product-overlapping standards (so-called generic standards) for verification of EMC compatibility.

The following changes are made to numbering when converting standards documents to European standards.

CENELEC nn → EN 50000 + nn
CISPR nn   → EN 55000 + nn
IEC nn   → EN 60000 + nn

For the CENELEC area, the standards produced for EMC are based on a hierarchy of basic standards, generic standards and product standards, as follows:

Basic standards These describe phenomena-related measuring and test methods for verification of the EMC. Specifications for measuring instruments and test set-ups, such as the recommendation of interference immunity test levels, are also contained, but are only incorporated as binding limits in generic standards or product standards.

Generic standards These contain important general limits for the assessment of products for which no product-specific standards are available. For the EMC environment there are differences between the industrial field (EN standards contain the extension –2) and the environment of light industry, of trade and business and of residential areas (EN standards contain the extension –1).

Product and product family standards These describe specific environmental conditions and take precedence over generic standards. Limits are stipulated in generic standards with test methods and procedures being mainly specified for product families. In addition to the standards, there are still various recommendations, such as from the VDEW, which should be regarded as transitional solutions for the areas in which there are no product standards. The following examples of VDEW recommendations should be mentioned in the context of system perturbations.

- Basic principles for the assessment of system perturbations.
- Recommendation for digital station control.
- Recommendations for the avoidance of impermissible perturbations on audio-frequency telecontrol.

In the international field, standardisation is controlled by the IEC, which has prepared an extensive series of standards on the subject of system perturbations. This series has been partly converted into national standards by translation of the relevant IEC publications.

The IEC 1000 series of standards covers all the areas of electromagnetic compatibility. A distinction is made between conducted disturbances (frequency range up to a few tens of kilohertz) and non-conducted disturbances in the higher frequency range.

The clear structuring of EMC standards in IEC 1000 is shown in the following overview, with the further subdivision in this case concentrating on low-frequency functions. In the VDE classification, the corresponding specifications are given mainly as sections under classification numbers 0839 to 0847.

IEC 1000–1	Overview of the series of standards, definitions
IEC 1000–2	Compatibility levels, description of environment
−1	Description of phenomena
−2	Compatibility levels for public low voltage systems
−4	Compatibility levels for industrial systems
−5	Classification of the EMC environment
−6	Recommendations for low-frequency emitted interference in industrial systems
−7	Low-frequency magnetic fields
−12	Compatibility levels for public medium voltage systems
IEC 1000–3	Emitted interference limits for voltage fluctuations, harmonics and flicker
−1	General overview
−2	Limits for harmonic currents $I_1 \leq 16$ A
−3	Limits for voltage fluctuations and flicker $I_1 \leq 16$ A
−4	Limits for harmonic currents $I_1 > 16$ A
−5	Limits for voltage fluctuations and flicker $I_1 > 16$ A
−6	Limits for harmonic currents in the medium voltage and high - voltage ranges
−7	Limits for voltage fluctuations and flicker in the medium - voltage and high voltage ranges
IEC 1000–4	Methods of testing for emitted interference and interference immunity
−1	General overview
−7	Recommendations for measurement of harmonics
−11	Interference immunity from voltage sags and interruptions
−13	Interference immunity from harmonics and interharmonics

-14   Interference immunity from voltage fluctuations, asymmetry and frequency deviations

-15   Function description of flicker meter

-16   Conducted continuous disturbances ($f = 0 \ldots 150$ kHz)

IEC 1000–5   Description of remedial measures

IEC 1000–6   Interference immunity requirements, emitted interference limits

IEC 1000–9   Miscellaneous

The layout of VDE standards (in this case giving the VDE classification number) for electrical power supply is generally as follows.

VDE 0838   Perturbations in power supply systems
Part 1   General, definitions
Part 2   Harmonics
Part 3   Voltage fluctuations

VDE 0839   Electromagnetic compatibility
Part 2–2   Compatibility levels in public low voltage systems
Part 2–4   Compatibility levels in industrial systems
Part 6–2   Interference immunity, industrial areas
Part 10   Assessment of interference immunity
Part 81–1   Emitted interference; residential areas, light industry
Part 81–2   Emitted interference; industrial areas
Part 82–1   Interference immunity; residential areas, light industry
Part 88   Compatibility levels in public medium voltage systems
Part 160   Features of voltage in public systems
Part 217   Measurements of emitted interference at installation site

VDE 0843   Electromagnetic compatibility of measuring and control equipment in industrial metrology
Part 1
Part 2   Interference immunity from the discharge of static electricity
Part 3   Interference immunity from electromagnetic fields
Part 5   Interference immunity from impulse voltage
Part 6   Interference immunity from conducted disturbances (HF fields)
Part 20
EMC requirements for electrical equipment for instrumentation and control and for use in laboratories
Part 23

VDE 0845   Protection of telecommunication systems from lightning, static charging and overvoltages from power systems

VDE 0846   Measuring devices for assessment of EMC
Part 0   Flicker meters, assessment of flicker strength
Part 1   Harmonics up to 2500 Hz

1.4 Basic mathematical principles

1.4.1 Complex calculations, vectors and phasor diagrams

When dealing with a.c. and three-phase systems, it should be noted that currents and voltages are generally not in phase. The phase position depends on the amount of inductance, capacitance and ohmic resistances at the impedance.

The time course, e.g. of a current or voltage in accordance with

$$u(t) = \sqrt{2}\, U \sin (\omega t + \varphi_{U}) \tag{1.1a}$$

$$i(t) = \sqrt{2}\, I \sin (\omega t + \varphi_{I}) \tag{1.1b}$$

can in this case be shown as a line diagram (see Figure 1.4). In the case of sinusoidal variables, these can be shown in the complex numerical level by rotating pointers, which rotate in the mathematically-positive sense (counterclockwise) with angular velocity ω as follows:

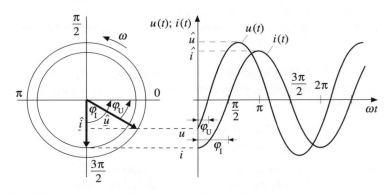

Figure 1.4 Vector diagram and time course of a.c. voltage

$$\underline{U} = \sqrt{2}\, U\, e^{(j\omega t + \varphi_U)} \tag{1.2a}$$

$$\underline{I} = \sqrt{2}\, I\, e^{(j\omega t + \varphi_i)} \tag{1.2b}$$

The time course in this case is obtained as a projection on to the real axis, see Figure 1.4.

DIN 40110 (VDE 0110) stipulates the terms for the designation of resistances and admittances. This specifies the following.

Resistance R Active resistance
Reactance X Reactance
Conductance G Active conductance
Susceptance B Susceptance

The generic term for resistances is given as impedance or apparent resistance

$$\underline{Z} = R + jX \tag{1.3a}$$

The generic term for conductances is admittance or apparent admittance

$$\underline{Y} = G + jB \tag{1.3b}$$

The reactance depends on the particular frequency under consideration and can be calculated for capacitances or inductances from

$$X_C = 1/\omega C \tag{1.4a}$$

$$X_L = \omega L \tag{1.4b}$$

For sinusoidal variables, the current through a capacitor, or the voltage at an inductance, can be calculated as follows.

$$i(t) = C \cdot du(t)/dt \tag{1.5a}$$

$$u(t) = L \cdot di(t)/dt \tag{1.5b}$$

The derivation for sinusoidal variables establishes that the current achieves, by an inductance, its maximum value a quarter period after the voltage. When considering the process in the complex level, the pointer for the voltage precedes the pointer for the current by 90°. This corresponds to a multiplication by +j.

For a capacitance, on the other hand, the voltage does not reach its maximum value until a quarter period after the current, the voltage pointer lags behind the current by 90°, which corresponds to a multiplication by –j.

This enables the relationships between current and voltage for inductances and capacitances to be shown in a complex notation.

$$\underline{U} = j\omega L \cdot \underline{I} \tag{1.6a}$$

$$\underline{I} = (1/(j\omega C)) \cdot \underline{U} \tag{1.6b}$$

Vectors are used to describe electrical processes. They are therefore used in d.c., a.c. and three-phase systems. Vector systems can, by definition, be chosen as required, but must not be changed during an analysis or calculation. It should also be noted that the appropriate choice of the vector system is of substantial assistance in describing and calculating special tasks. The need for vector systems is clear if one considers the Kirchhoff laws, for which the positive direction of currents and voltages must be specified. In this way, the positive directions of the active and reactive powers are then also stipulated.

For reasons of comparability and transferability, the vector system for the three-phase network (RYB components) should also be used for other component systems (e.g. symmetrical components), which describe the three-phase network.

If vectors are drawn as shown in Figure 1.5, the active and reactive powers, for instance output by a generator in overexcited operation, are positive. This vector system is designated as a generator vector system. Accordingly, the active and reactive power consumed by the load is positive when choosing the consumer vector system.

When describing electrical systems, voltage vectors are drawn from the phase conductor (L1, L2, L3 or also R, Y, B) to earth (E). In other component systems, for instance for a system of symmetrical components (see section 1.4.3), the voltage vector is shown from the conductor towards the particular reference rail. On the other hand, vectors in phasor diagrams are shown in the opposite direction. The vector of a conductor to earth voltage is therefore shown in the phasor diagram from earth potential to conductor potential.

Based on the stipulation of the vector system, the voltage and current relationship of an electrical system can be shown in phasor diagrams. Where steady-state or quasi-steady-state operation is shown, r.m.s. value phasors are generally used. Figure 1.6 shows the phasor diagram of an ohmic-inductive load in the generator and in the consumer vector system.

1.4.2 Fourier analysis and synthesis

The previously-considered processes in linear systems where currents and voltages occur with only one frequency can also be transferred to networks with any current or voltage characteristic. This is based on the known fact that any

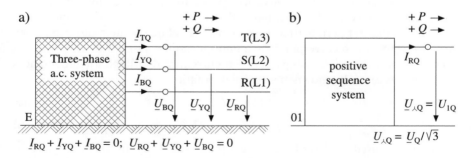

Figure 1.5   Definition of vectors for current, voltage and power in three-phase a.c. systems
a) power system diagram
b) electrical diagram for symmetrical conditions (positive sequence system)

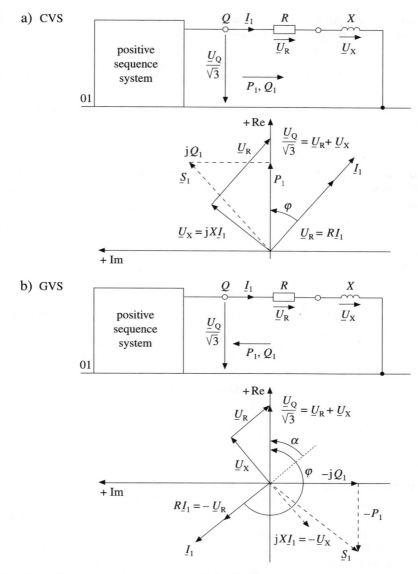

Figure 1.6 Vector diagram of current, voltage and power
  a) related to consumers (consumer vector system)
  b) related to power generation (generator vector system)

periodic signal with the period T can be represented by a Fourier series in accordance with the following equation:

$$f(t) = \frac{a_0}{2} + \sum_{h=1}^{\infty} (a_h \cos(h\omega_1 t) + b_h \sin(h\omega_1 t)) \tag{1.7}$$

The relationship between the period T and the basic circuit frequency ω_1 is obtained by

$$\omega_1 T = 2\pi \tag{1.8}$$

A particular simple representation of the Fourier coefficients which is adapted to the calculation with complex numbers is obtained by combining the coefficients a_h and b_h:

$$c_h = (a_h - jb_h) \tag{1.9a}$$

with the amplitude of the harmonic component c_h and phase position φ_h in accordance with

$$\varphi_h = \arctan(a_h/b_h) \tag{1.9b}$$

The content where $h = 1$ forms the fundamental component, the content where $h > 1$ forms the harmonics.

The coefficients a_h and b_h can be determined in accordance with

$$a_h = \frac{1}{\pi} \int_0^{2\pi} f(t) \cos(h\omega_1 t) \, d\omega t \tag{1.10a}$$

$$b_h = \frac{1}{\pi} \int_0^{2\pi} f(t) \sin(h\omega_1 t) \, d\omega t \tag{1.10b}$$

The integrals can generally only be assessed numerically.

Where a signal is sampled (periodically in 2π), the Fourier coefficients can be calculated approximately by summation. The functional course $f(t)$ shown in Figure 1.7 is given as an example.

For sampling in equidistant intervals (subdivisions of the period interval $0 \le x \le 2\pi$ to an uneven number $n = 2N + 1$ subintervals of length $l = 2\pi/n$) the approximate values a_h and b_h where $h = 0.1 \ldots, N$ is obtained for the Fourier coefficients in accordance with

$$a_h(l) = \frac{1}{\pi} \sum_{k=0}^{2N} f_k \cos(hkl) \tag{1.11a}$$

$$b_h(l) = \frac{1}{\pi} \sum_{k=0}^{2N} f_k \sin(hkl) \tag{1.11b}$$

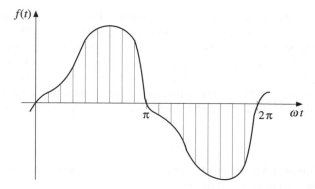

Figure 1.7   Time course of function $f(t)$; 18 samples per period

where $f_k = f(kl)$. We then get

$$f(x;l) = 0.5a_0 + \sum_{h=1}^{n} a_h(l) \cos(hx) + \sum_{h=1}^{N} b_h(l) \sin(hx) \qquad (1.12)$$

the trigonometric approximation polynomial of the Nth order for $f(x)$, which agrees with $f(x)$ at positions $x = hl$.

For subdivision of the interval into an even number $n = 2N$ of subintervals of length $l = 2\pi/n$, the element a_N is provided with the factor 0.5 to obtain the interpolation attribute. For the Fourier coefficients, the approximate values a_h and b_h for $h = 0.1 \ldots, N$ are obtained in accordance with

$$a_h(l) = \frac{1}{\pi} \sum_{k=0}^{2N-1} f_k \cos(hkl) \qquad (1.13a)$$

and for $h = 1, \ldots, N-1$ in accordance with

$$b_h(l) = \frac{1}{\pi} \sum_{k=1}^{2N-1} f_k \sin(hkl) \qquad (1.13b)$$

The trigonometric approximation polynomial is then

$$f(x;l) = 0.5a_0 + \sum_{h=1}^{2N-1} a_h(l) \cos(hx) + \sum_{h=1}^{2N-1} b_h(L) \sin(hx) +$$

$$+ 0.5a_N(l) \cos(hx) \qquad (1.14)$$

In this case, h signifies the order of the harmonic and $n = 2N$ or $n = 2N + 1$ the number of sampling values per period of the fundamental component. The equations show that to represent the content with the order h of the harmonic, at least the number of $2h$ sampling values per period of fundamental frequency are required. From this it follows that with a fixed signal sampling frequency the highest harmonic that can be represented is that with half the frequency of the sampled signal (Shannon sampling theorem).

To determine the Fourier coefficients (discrete Fourier transformation), various methods such as direct calculation, prime factor algorithm or butterfly algorithm can be used. The importance of choosing a mathematical method is reduced with the availability of signal processors. These provide the Fourier coefficients required for further processing from the sampled values.

If the analysing function has symmetrical properties, the calculation of the Fourier coefficients is substantially simplified [4]. Assuming that the function $f(t)$ is odd, i.e.

$$f(t) = -f(-t) \qquad (1.15)$$

(see also Figure 1.8), all coefficients c_h become purely imaginary and the Fourier coefficient a_h is equal to zero.

In the case of a function $f(t)$ which is odd with the half period in accordance with Figure 1.8, the following applies.

$$f(t) = -f(t - (T/2)) \qquad (1.16)$$

In this case, all even-numbered coefficients become zero. If the function is $f(t)$, e.g. the voltage $u(t)$ at a non-linear resistor through which the sinusoidal current $i(t)$ flows, the voltage has odd-order harmonics. This characteristic is known as

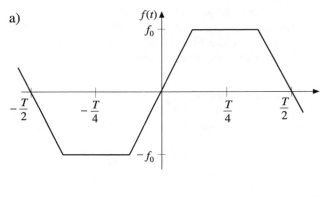

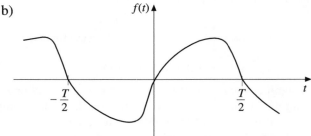

Figure 1.8   Time course function f(t)
a) even function
b) odd function (symmetrical to half period)

central-symmetric. Such characteristics occur frequently in electrical power supply systems.

1.4.3 Symmetrical components

The relationships between voltages and currents of a three-phase system can be represented by a matrix equation, e.g. with the aid of the impedance or admittance matrix. The equivalent circuits created by electrical equipment, such as lines, cables, transformers and machines, in this case have couplings in the three-phase system which are of an inductive, capacitive and galvanic type. This can be explained by using any short element of an overhead line in accordance with Figure 1.9 as an example, see also [2].

The correlation of currents and voltages of the RYB system is as follows:

$$\begin{bmatrix} \underline{U}_R \\ \underline{U}_Y \\ \underline{U}_B \end{bmatrix} = \begin{bmatrix} \underline{Z}_{RR} & \underline{Z}_{RY} & \underline{Z}_{RB} \\ \underline{Z}_{YR} & \underline{Z}_{YY} & \underline{Z}_{YB} \\ \underline{Z}_{BR} & \underline{Z}_{BY} & \underline{Z}_{BB} \end{bmatrix} \cdot \begin{bmatrix} \underline{I}_R \\ \underline{I}_Y \\ \underline{I}_B \end{bmatrix} \tag{1.17}$$

All the values of this impedance matrix can generally be different. Because of the cyclic-symmetrical construction of three-phase systems only the self-impedance and two coupling impedances are to be considered. A cyclic-symmetrical matrix is thus obtained.

$$\begin{bmatrix} \underline{U}_R \\ \underline{U}_Y \\ \underline{U}_B \end{bmatrix} = \begin{bmatrix} \underline{Z}_A & \underline{Z}_B & \underline{Z}_C \\ \underline{Z}_C & \underline{Z}_A & \underline{Z}_B \\ \underline{Z}_B & \underline{Z}_C & \underline{Z}_A \end{bmatrix} \cdot \begin{bmatrix} \underline{I}_R \\ \underline{I}_Y \\ \underline{I}_B \end{bmatrix} \tag{1.18}$$

The multiplicity of couplings between the individual components of three-phase systems complicates the solution methods, particularly when calculating extended networks. For this reason, a mathematical transformation is sought

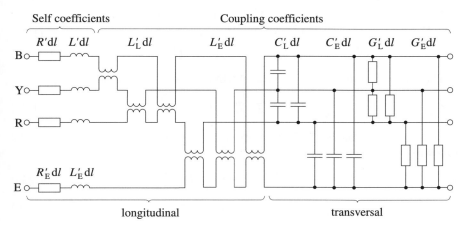

Figure 1.9   Differentially small section of homogeneous three-phase a.c. line

which transfers the RYB components to a different system. The following conditions should apply for the transformation.

- The transformed voltages should depend only on one transformed current.
- For symmetrical operations only one component should be unequal to zero.
- The linear relationship between current and voltage should be retained, i.e. the transformation should be linear.
- For symmetrical operations the current and voltage of the reference component should be retained (reference component invariant).

The desired transformation should, in this case, enable the three systems to be decoupled in such a way that the three components are decoupled from each other in the following manner:

$$\begin{bmatrix} U_0 \\ U_1 \\ U_2 \end{bmatrix} = \begin{bmatrix} Z_0 & 0 & 0 \\ 0 & Z_1 & 0 \\ 0 & 0 & Z_2 \end{bmatrix} \cdot \begin{bmatrix} I_0 \\ I_1 \\ I_2 \end{bmatrix} \tag{1.19}$$

These requirements are fulfilled by the transformation to the symmetrical components (012-system), which is realised for voltages and currents by the transformation matrix $\underline{T}$ according to Equation (1.20), shown for the voltages. It should be noted that the factor 1/3 is part of the transformation and therefore belongs to the matrix $\underline{T}$.

$$\begin{bmatrix} U_0 \\ U_1 \\ U_2 \end{bmatrix} = \frac{1}{3}\begin{bmatrix} 1 & 1 & 1 \\ 1 & \underline{a} & \underline{a}^2 \\ 1 & \underline{a}^2 & \underline{a} \end{bmatrix} \cdot \begin{bmatrix} U_R \\ U_Y \\ U_B \end{bmatrix} \tag{1.20}$$

The reverse transformation of the 012 system to the RYB system is achieved by the matrix $\underline{T}^{-1}$ in accordance with the following equation:

$$\begin{bmatrix} U_R \\ U_Y \\ U_B \end{bmatrix} = \frac{1}{3}\begin{bmatrix} 1 & 1 & 1 \\ 1 & \underline{a}^2 & \underline{a} \\ 1 & \underline{a} & \underline{a}^2 \end{bmatrix} \cdot \begin{bmatrix} U_0 \\ U_1 \\ U_2 \end{bmatrix} \tag{1.21}$$

The following applies for both transformation matrices $\underline{T}$ and $\underline{T}^{-1}$

$$\underline{T} \cdot \underline{T}^{-1} = \underline{E} \tag{1.22}$$

with the identity matrix $\underline{E}$. The complex rotational phasors $\underline{a}$ and $\underline{a}^2$ have the following meanings:

$$\underline{a} = e^{j120°} = -1/2 + j(1/2)\sqrt{3} \tag{1.23a}$$

$$\underline{a}^2 = e^{j240°} = -1/2 - j(1/2)\sqrt{3} \tag{1.23b}$$

$$1 + \underline{a} + \underline{a}^2 = 0 \tag{1.23c}$$

For the transformation of the impedance matrix, Equation (1.24) applies in accordance with the laws of matrix multiplication, taking account of Equations (1.20) and (1.22).

$$\boldsymbol{T}\,\underline{U}_{RYB} = \boldsymbol{T}\,\underline{\boldsymbol{Z}}_{RYB}\boldsymbol{T}^{-1}\boldsymbol{T}\,\underline{I}_{RYB} \tag{1.24a}$$

$$\underline{U}_{012} = \underline{\boldsymbol{Z}}_{012}\,\underline{I}_{012} \tag{1.24b}$$

and thus Equation (1.25) for the conversion of the impedances of the three-phase system to the 012 system.

$$\underline{Z}_0 = \underline{Z}_A + \underline{Z}_B + \underline{Z}_C \tag{1.25a}$$

$$\underline{Z}_1 = \underline{Z}_A + \underline{a}^2\underline{Z}_B + \underline{a}\,\underline{Z}_C \tag{1.25b}$$

$$\underline{Z}_2 = \underline{Z}_A + \underline{a}\,\underline{Z}_B + \underline{a}^2\underline{Z}_C \tag{1.25c}$$

The impedance values of the positive sequence and negative sequence systems are generally equal. This applies to all non-rotating equipment. The zero sequence impedance mainly has a different value from the positive or negative sequence impedance. If mutual coupling is absent, as perhaps with three single-pole transformers connected together to form a three-phase transformer, the zero sequence impedance is equal to the positive or negative sequence impedance.

The voltage vector of the RYB system is linked linearly to the voltage vector of the 012 system (the same applies for the currents).

If only one zero sequence system exists, the following applies:

$$\begin{bmatrix} \underline{U}_R \\ \underline{U}_Y \\ \underline{U}_B \end{bmatrix} = \begin{bmatrix} 1 & 1 & 1 \\ 1 & \underline{a}^2 & \underline{a} \\ 1 & \underline{a} & \underline{a}^2 \end{bmatrix} \cdot \begin{bmatrix} \underline{U}_0 \\ 0 \\ 0 \end{bmatrix} = \begin{bmatrix} \underline{U}_0 \\ \underline{U}_0 \\ \underline{U}_0 \end{bmatrix} \tag{1.26}$$

No phase shift exists between the three a.c. systems of the **RYB** conductors. The zero sequence system is thus an a.c. system. Figure 1.10 shows the phasor diagram of the voltages of the **RYB** system and the voltage of the zero sequence system.

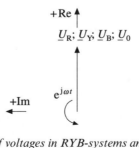

Figure 1.10   Vector diagram of voltages in RYB-systems and zero sequence systems
Positive and negative sequence systems are NIL

Where only a positive sequence exists, the following applies:

$$
\begin{bmatrix} \underline{U}_R \\ \underline{U}_Y \\ \underline{U}_B \end{bmatrix} = \begin{bmatrix} 1 & 1 & 1 \\ 1 & \underline{a}^2 & \underline{a} \\ 1 & \underline{a} & \underline{a}^2 \end{bmatrix} \cdot \begin{bmatrix} 0 \\ \underline{U}_1 \\ 0 \end{bmatrix} = \begin{bmatrix} \underline{U}_1 \\ \underline{a}^2 \underline{U}_1 \\ \underline{a}\,\underline{U}_1 \end{bmatrix}
\tag{1.27}
$$

A three-phase system with a positive rotating phase sequence R, Y, B results, i.e. a positive sequence system. Figure 1.11 shows the phasor diagram of the voltages of the RYB system and the voltage of the positive sequence system.

Where only a negative sequence system exists, the following applies.

$$
\begin{bmatrix} \underline{U}_R \\ \underline{U}_Y \\ \underline{U}_B \end{bmatrix} = \begin{bmatrix} 1 & 1 & 1 \\ 1 & \underline{a}^2 & \underline{a} \\ 1 & \underline{a} & \underline{a}^2 \end{bmatrix} \cdot \begin{bmatrix} 0 \\ 0 \\ \underline{U}_2 \end{bmatrix} = \begin{bmatrix} \underline{U}_2 \\ \underline{a}\,\underline{U}_2 \\ \underline{a}^2 \underline{U}_2 \end{bmatrix}
\tag{1.28}
$$

A three-phase system with a positive counterrotating phase sequence R, B, Y results, i.e. a negative sequence system. Figure 1.12 shows the phasor diagram of the voltages of the RYB system and the voltage of the negative sequence system.

Three-phase networks in cyclic terms are generally symmetrically constructed and operated. This must therefore also apply to currents with a harmonic

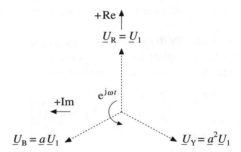

Figure 1.11   Vector diagram of voltages in RYB-systems and positive sequence systems
Zero and negative sequence systems are NIL

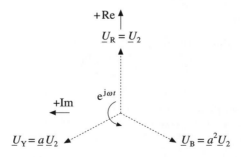

Figure 1.12   Vector diagram of voltages in RYB-systems and negative sequence systems
Zero and positive sequence systems are NIL

component [3]. If the current $i_R(t)$ is represented by a Fourier series in accordance with Equation (1.29a),

$$i_R(t) = \sum_{h=1}^{\infty} \sqrt{2} I_h \sin{(h\omega t + \varphi_{ih})} \tag{1.29a}$$

then based on the general correlation

$$i_Y(t) = i_R(t - (T/3)) \tag{1.29b}$$

$$i_B(t) = i_R(t + (T/3)) \tag{1.29c}$$

we get the currents $i_Y(t)$ and $i_B(t)$

$$i_Y(t) = \sum_{h=1}^{\infty} \sqrt{2}\, I_h \sin{(h\omega t - h2\pi/3 + \varphi_{ih})} \tag{1.30a}$$

$$i_Y(t) = \sum_{h=1}^{\infty} \sqrt{2}\, I_h \sin{(h\omega t + h2\pi/3 + \varphi_{ih})} \tag{1.30b}$$

Therefore the phase angle $\varphi = \pm h2\pi/3$ between conductors R, Y and B. Where the symmetry of the phase-variables is cyclic, the harmonics form positive sequence, negative sequence and zero sequence (homopolar components) in accordance with their order ($h = 0, 1, 2, 3 \ldots$), as follows:

$3h + 1$	Positive sequence system
$3h + 2$	Negative sequence system
$3h$	Zero sequence system

In symmetrically-constructed three-phase networks, the currents of the zero sequence system flow via the neutral conductor and earth at three times the value. Where there is no neutral point earthing, a harmonic component of the voltage of the neutral point with respect to earth forms for the corresponding frequency.

1.4.4 Power considerations

The instantaneous value of the power $p(t)$ in an a.c. circuit is calculated as

$$p(t) = u(t)\, i(t) \tag{1.31}$$

with the instantaneous values of the current $i(t)$ and of the voltage $u(t)$. Generally, this product has positive and negative values during a period. The mean power in accordance with Equation (1.32) is termed the active power.

$$P = \frac{1}{T} \int_0^T u(t) i(t) \mathrm{d}t \tag{1.32}$$

If sinusoidal current and sinusoidal voltage are assumed, i.e.

$$u(t) = \sqrt{2}\, U \cos (\omega t + \varphi_U), \tag{1.33a}$$

$$i(t) = \sqrt{2}\, I \cos (\omega t + \varphi_I), \tag{1.33b}$$

then as a product of the instantaneous values of the current and voltage the following

$$p(t) = 2UI \cos (\omega t + \varphi_U) \cos (\omega t + \varphi_I) \tag{1.34a}$$

$$p(t) = UI \cos \varphi + UI \cos (2\omega t + \varphi) \tag{1.34b}$$

apply as the instantaneous values of the power where $\varphi = \varphi_U + \varphi_I$. The power $p(t)$ oscillates about the mean value $UI \cos \varphi$ at twice the frequency. This mean value is the active power P. The product UI is designated the apparent power S.

If φ_U or φ_I in Equation (1.34) is eliminated, we get:

$$p(t) = UI \cos \varphi + UI \cos (2\omega t + 2\varphi_I) - UI \sin \varphi \sin (2\omega t + 2\varphi_I) \quad (1.35a)$$

$$p(t) = UI \cos \varphi + UI \cos (2\omega t + 2\varphi_U) + UI \sin \varphi \sin (2\omega t + 2\varphi_U) \quad (1.35b)$$

The variable $UI \sin \varphi$ is designated the reactive power Q. It also oscillates at twice the frequency, but about the zero-frequency mean value. The reactive power is positive if the angle φ is between $0°$ and $+180°$, i.e. if the voltage leads the current.

In each case the following applies.

$$|Q| = \sqrt{S^2 - P^2} \tag{1.36}$$

The quotient of the active power P and apparent power S is called the power factor $\cos \varphi$.

In the case of non-sinusoidal currents and voltages, described by the sum of the fundamental component and harmonics in accordance with the results of the Fourier analysis, it should be noted that currents and voltages can convert only active power if they are of equal frequency, because the integral for currents and voltages of unequal frequency in accordance with Equation (1.32) makes no contribution.

$$P = \frac{1}{T} \int_0^T u(t)i(t) \, dt \tag{1.32}$$

If with non-sinusoidal currents and voltages Equation (1.37) is used, i.e.

$$u(t) = \sum_{k=1}^{N} \sqrt{2} U_k \cos (k\omega_1 t + \varphi_{Uk}), \tag{1.37a}$$

$$i(t) = \sum_{l=1}^{N} \sqrt{2} I_l \cos (l\omega_1 t + \varphi_{Il}), \tag{1.37b}$$

the instantaneous value of the power is calculated as

$$p(t) = \sum_{k=l=1}^{N} 2U_k I_l \cos (\varphi_{Uk} - \varphi_{Il}) +$$

$$+ \sum_{k=1}^{N} \sum_{l=1}^{N} U_k I_l \cos ((k+l)\omega_1 t + \varphi_{Uk} + \varphi_{Il}) +$$

$$+ \sum_{\substack{k=1 \\ k \neq l}}^{N} \sum_{l=1}^{N} U_k I_l \cos ((k-l)\omega_1 t + \varphi_{Uk} - \varphi_{Il}) \qquad (1.38)$$

The first summand describes the active power, whereby the component with $k = l = 1$ represents the fundamental component active power. The summands where $k = l > 1$ render the harmonic active powers. The second summand renders the reactive power Q and the third summand the distortive power D. The time course of these powers oscillates non-sinusoidally about the zero-frequency mean value. The course of the powers can also be shown in more complex representations as phasors. The time course is then the projection of the rotating phasor on the real axis in accordance with section 1.4.1. Between the powers, the correlations

$$S^2 = P^2 + Q^2 + D^2, \qquad (1.39)$$

apply, which can also be represented in diagram form, as shown in Figure 1.13.

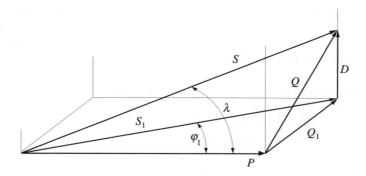

Figure 1.13   Vector diagram for the definition of different types of power in a.c. systems according to DIN 40110

1.4.5 Series and parallel resonant circuits

To analyse electrical networks, e.g. an electrical power supply network, it is necessary to calculate series and parallel circuits of equipment.

If capacitances, inductances and resistances are present in this network, the

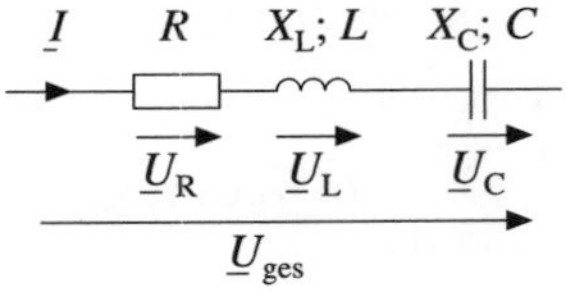

Figure 1.14   Electrical diagram of a series resonance circuit

corresponding circuit represents a series or parallel resonant circuit. Such arrangements are frequently found in electrical power supply networks and it must be possible to analyse their behaviour where there are high-frequency components in the current and voltage.

The series resonant circuit shown in Figure 1.14 is considered first [4].

The impedance of the series resonant circuit is calculated as follows:

$$\underline{Z} = R + j\omega L - j\frac{1}{\omega C} \qquad (1.40)$$

Resonance is present if the imaginary part of the impedance $\underline{Z}$ becomes zero. This is the case for the resonant circuit frequency ω_{res} according to the following equation.

$$\omega_{res} = \frac{1}{\sqrt{LC}} \qquad (1.41)$$

At resonant frequency the impedance of the series resonant circuit is very small, and is limited only by the value of the resistance R. For frequencies ω above the resonant frequency ω_{res} the impedance of the series resonant circuit becomes inductive and at frequencies ω below the resonant frequency ω_{res} the impedance is capacitive. When a voltage is applied to the resonant circuit, the current through the resonant circuit increases as the frequency approaches the resonant frequency.

The course of the amount of the impedance of the resonant circuit is shown in Figure 1.15.

The relation shown in Equation (1.42) is called the attenuation A (sometimes named d) of the series resonant circuit.

$$d = R\sqrt{\frac{C}{L}} \qquad (1.42)$$

The reciprocal of the attenuation is called the quality Q. A further variable for describing a resonant circuit is the bandwidth B. It is defined by two frequencies (ω_+ and ω_-) above and below the resonant frequency ω_{res} at which the amount of the impedance $\underline{Z}$ has risen to $\sqrt{2}$ – times the value, relative to the impedance value at resonant frequency (see Figure 1.15). The following applies for the bandwidth.

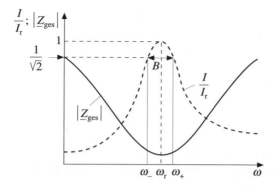

Figure 1.15 Impedance versus frequency of a series resonance circuit (see also Figure 1.14)

$$\omega_+ - \omega_- = R/L \qquad (1.43)$$

The voltage of the individual components of the series resonant circuit increases as the resonant frequency is approached. The voltages in this case are calculated as:

$$U_L = \frac{j\omega L}{Z} U \qquad (1.44a)$$

$$U_C = \frac{1}{j\omega CZ} U \qquad (1.44b)$$

By conversion and reference to the applied total voltage U, we get:

$$\frac{U_L}{U} = \frac{\omega/\omega_{res}}{\sqrt{d^2 + (\omega/\omega_{res} - \omega_{res}/\omega)^2}} \qquad (1.45a)$$

$$\frac{U_C}{U} = \frac{\omega_{res}/\omega}{\sqrt{d^2 + (\omega/\omega_{res} - \omega_{res}/\omega)^2}} \qquad (1.45b)$$

The magnitude of the voltage U_L at the inductivity or U_C at the capacitor is, under certain circumstances, in the vicinity of the resonant frequency substantially greater than the amount of the total voltage U, depending on the quality of the resonant circuit.

For the parallel resonant circuit shown in Figure 1.16 the relationships are similar.

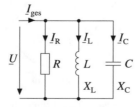

Figure 1.16 Electrical diagram of a parallel resonance circuit

The admittance of the parallel resonant circuit is calculated as follows:

$$\underline{Y} = \frac{1}{R} + j\omega C - j\frac{1}{\omega L} \tag{1.46}$$

At the resonant circuit frequency

$$\omega_{\text{res}} = \frac{1}{\sqrt{LC}} \tag{1.47}$$

the imaginary part of the admittance $\underline{Y}$ becomes zero. The impedance of the parallel resonant circuit at resonant frequency is very large and is limited only by the value of the resistance R.

For frequencies ω above the resonant frequency ω_{res} the impedance of the series resonant circuit is inductive and for frequencies ω below the resonant frequency ω_{res} the impedance is capacitive. If a current flows through the resonant circuit, the voltage on the resonant circuit increases as the frequency approaches the resonant frequency. The course of the magnitude of the impedance of the parallel resonant circuit is shown in Figure 1.17.

Attenuation d and quality Q of the parallel resonant circuit are defined in a similar way to the series resonant circuit.

$$d = \frac{1}{R}\sqrt{\frac{L}{C}} \tag{1.48}$$

The bandwidth of the parallel resonant circuit is defined by the two frequencies ω_+ and ω_- above and below the resonant frequency ω_{res}, at which the magnitude of the admittance $\underline{Y}$ has risen to the $\sqrt{2}$-times value, relative to the value of the admittance at resonant frequency (see Figure 1.17) and the impedance has thus dropped to the $\sqrt{2}$-times value. The bandwidth B is calculated as follows:

$$\omega_+ - \omega_- = 1/RC \tag{1.49}$$

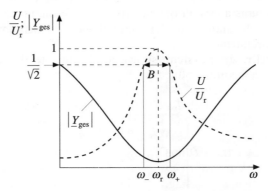

Figure 1.17   Impedance versus frequency of a parallel resonance circuit (see also Figure 1.16)

The current through the individual components of the parallel resonant circuit increases as the resonant frequency is approached. The currents in this case are calculated as follows.

$$I_L = \frac{1}{j\omega LY} I \qquad (1.50a)$$

$$I_C = \frac{j\omega C}{Y} I \qquad (1.50b)$$

By conversion and reference to the total current I we get the following

$$\frac{I_C}{I} = \frac{\omega/\omega_{res}}{\sqrt{d^2 + (\omega/\omega_{res} - \omega_{res}/\omega)^2}} \qquad (1.51a)$$

$$\frac{I_L}{I} = \frac{\omega_{res}/\omega}{\sqrt{d^2 + (\omega/\omega_{res} - \omega_{res}/\omega)^2}} \qquad (1.51b)$$

The amount of the current I_L due to the inductivity or I_C through the capacitor is, under certain circumstances, substantially greater than the amount of the total current I, depending on the quality of the resonant circuit in the vicinity of the resonant frequency.

1.5 System conditions

1.5.1 Voltage levels and impedances

When considering system perturbations it is necessary to include the impedance of the feeding network because, for instance, a non-sinusoidal current at the impedance of the infeed causes a non-sinusoidal voltage drop. System perturbations in this case occur at all voltage levels of power supply systems. The level of the disturbance phenomenon is in this case dependent on the ratio of the system impedances.

Figure 1.18 shows a simplified arrangement of the basic design of the electrical power supply system with respect to the treatment of system perturbations. In this case, harmonics are considered as disturbance phenomena.

In examining the three system levels (380 kV and 110 kV as a high voltage system, 10 kV as a medium voltage system and 0.4 kV as a low voltage system) it is assumed that power stations prefer to supply at the 380 kV level. Power station supplies at other system levels do not change the basic method of examination.

In the 0.4 kV system, a source of harmonics is assumed which feeds any harmonic spectrum $\underline{I}_{hLV}$ into the system. These currents cause voltage drops $\underline{U}_{hLV}$ at the impedance of the feeding transformer and at the series system impedances. The 10/0.4 kV transformer is connected by one, or more, lines, to a 10 kV network node. Other low voltage systems can be connected here by lines and transformers, and large industrial consumers or powerful harmonic sources are also possible here.

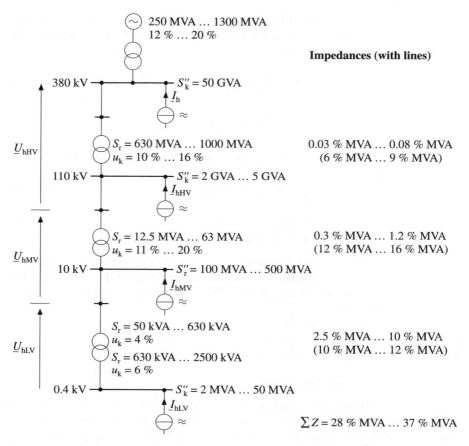

Figure 1.18   Principal diagram of electrical system with different voltage levels with respect to perturbations, e.g. harmonics

It should therefore be assumed that harmonic currents $\underline{I}_{hMV}$ are injected into the 10 kV system at the 10 kV network node. These currents are superimposed with regard to their angular position by the currents fed in from the 0.4 kV system and lead to voltage drops $\underline{U}_{hMV}$ at the impedance of the feeding 110/10 kV transformer and at the series impedances. This is repeated at the 110 kV level. The harmonic currents are thus superimposed from the lower to the higher voltage levels, while the voltage drops from the higher system levels also act on the lower system levels—the voltage drops are carried over from the higher to the lower voltage level.

If typical values for equipment, as shown in Figure 1.18, i.e. the rated values of impedance voltage and apparent power of the transformers as well as the impedances of the system, supply lines at the individual voltage levels, are considered, it can be seen that the initial symmetrical short-circuit power of the

Table 1.1   Details of typical symmetrical short-circuit power in systems

U_n in kV	S''_k in GVA
HV: 380	50
HV: 110	2 . . . 5
MV: 10	0.1 . . . 0.5
LV: 0.4	0.02 . . . 0.05

individual voltage levels are smaller by approximately one order of magnitude, as shown in Table 1.1.

The impedance ratios of the three system levels then amount to the following.

$$z_{HV}:z_{MV}:z_{LV} = 6 \ldots 9\%/\text{MVA}: 12 \ldots 16\%/\text{MVA}: 10 \ldots 12\%/\text{MVA}$$

1.5.2 Features of the voltage in power systems

The earliest attempts to specify the parameters of voltage date from 1989, by UNIPEDE, which described the actual status in low and medium voltage systems. On the basis of this document, CENELEC passed a European standard EN 50160 in 1993, which described the features of voltage and frequency in public supply systems. This standard has been in force since October 1995.

It contains a description of the essential features of the voltage in public supply systems at the customer connection point (point of common coupling). High voltage systems are not considered. The features of the voltage are not intended as values of electromagnetic compatibility or as conducted emitted interference limits. EN 50160 specifies no electrotechnical safety regulations going beyond this and therefore has no VDE classification number.

The features of the supply voltage for low voltage systems are given in the following.

- Power frequency variation
  Ten-second mean value of fundamental frequency
    50 Hz ± 1% during 95% of one week
    50 Hz + 4%/−6% during 100% of one week

- Voltage level
  Low voltage, three-wire, three-phase systems
    U_n = 230 V between phase conductors
  Low voltage, four-wire, three-phase systems
    U_n = 230 V between phase conductor and neutral conductor
    (up to the year 2003 the voltage band can deviate from this in accordance with HD 472 S1)

- Slow voltage variations
  95% of the ten-minute mean values of the system voltage
    $U = U_n \pm 10\%$

(up to the year 2003 the voltage band can deviate from this in accordance
with HD 472 S1)

- Fast voltage changes
  $\Delta u \leq 5\%$ (up to 10% for short duration several times a day)
    A voltage change of > 10% is defined as a voltage sag

- Flicker
  $P_{lt} \leq 1$ for 95% of the week

- Voltage sags
  $\Delta u \leq 40\%$; $t < 1$ s
    $n = 10$ to 1000 per year, with isolated sags also being of greater duration,
    depth and frequency

- Short time interruptions
  $n = 10$ to 500 per year
    $t < 1$s for 70% of all interruptions
    Design of protective devices up to 3 min

- Long time interruptions
  $n = 10$ to 50 per year
    $t > 3$ min

- Temporary (power system frequency) overvoltages
  $U_{max} = 1.5$ kV between phase conductor and earth for short-circuits on the
  high voltage side of a transformer

- Transient overvoltages
  $U_{max} < 6$ kV
    Rise times in the microsecond range; the energy content of the overvoltage
    is determinant for the effect

- Voltage unbalance
  95% of the ten-minute r.m.s. values of a week
    $U_{neg} \leq 0.02 \ U_{pos}$
  Exception for many a.c. consumers
    $U_{neg} \leq 0.03 \ U_{pos}$

- Total harmonic distortion
  95% of the ten-minute means values of stated table values;
  total harmonic distortion (THD) to $h = 40$: $THD \leq 8\%$

- Interharmonics
  No data

It should be noted that the tabulated values of the harmonics in EN 50160
correspond to those of EN 61000–2–2 but are given only up to order $h = 25$.

The features of the supply voltage normally change within the stated limits.
There is, however, a certain probability that features can occur outside the stated

limits. It can therefore not be deduced from EN 50160 that the stated values and frequencies cannot be exceeded for individual customers or in certain parts of the system. The informative annex EN 50160 states the following.

'This standard stipulates, for the phenomena for which it is possible, the value ranges normally to be expected in which the features of the supply voltage change. For the other features, the standard provides the best possible guide values to be expected in systems.
. . .
Although this standard clearly has references to compatibility levels, it is important to expressly point out that this standard refers to electrical energy with regard to the features of the supply voltage. It is not a standard for compatibility levels.'

1.5.3 Impedances of equipment

It is necessary to calculate the values of the equipment of electrical supply systems in order, for instance, to examine the behaviour of the supply system during normal operation (power frequency load flow calculations), in the disturbed operating state (short-circuit current calculations) and for higher frequency occurrences (harmonics). In this connection, equipment such as generators, transformers, lines, motors and capacitors are of interest. Simulation of consumers is only necessary in special cases. The calculation of equipment data from name plate data or tabulated data is preferred. Various systems of units are available for calculation.

Physical quantities To describe the steady-state conditions of equipment and of the system requires four units, i.e. voltage U, current I, impedance Z and power S, with the units Volt, Ampere, Ohm and Watt, which are linked to each other by Ohm's law and the power equation.

If physical quantities are taken to be measurable properties of physical objects, occurrences and states from which useful sums and differences can be formed, the following then applies:

$$\text{Quantity} = \text{numerical value} \times \text{unit}$$

Relative quantities On the contrary, the unit of a relative quantity is by definition unity, i.e.

$$\text{Relative quantity} = \text{quantity/reference quantity}$$

Because the four quantities, voltage, current, impedance and power, required for system calculations are linked to each other, two reference quantities are required to specify a relative system of units. Voltage and power are usually chosen for this purpose. This provides the per-unit system which is widespread in the English-speaking world.

Semirelative quantities In the semirelative system of units only one quantity is freely chosen as the reference quantity. If the voltage U_B is chosen for this, the

%/MVA system is obtained, which is outstandingly suitable for network calculations because the values of the equipment can be very easily calculated. Table 1.2 gives the definitions in the various units. A conversion between the system is made using the data in Table 1.3.

The impedances or reactances for electrical equipment are determined from the data of the name plate or from geometrical dimensions. The reactances, resistances or impedances should generally be calculated relative to the nominal apparent power or the nominal voltage of the system in which the equipment is fitted.

Where the rated transformer ratios do not coincide with the system nominal voltages, correction factors must be considered [2].

Table 1.2   Definitions of quantities in physical, relative and semirelative units

Ohm system physical units	%/MVA system semirelative units	Per unit system relative units
No reference quantity	One reference quantity	Two reference quantities
Voltage U	$u = \dfrac{U}{U_B} = \dfrac{\{U\}}{\{U_B\}} \cdot 100\%$	$'u = \dfrac{U}{U_B} = \dfrac{\{U\}}{\{U_B\}} \cdot 1$
Current U	$i = I \cdot U_B = \{I\} \cdot \{U_B\} \cdot \text{MVA}$	$'i = \dfrac{I \cdot U_B}{S_B} = \dfrac{\{I\} \cdot \{U_B\}}{[S_B]} \cdot 1$
Impedance Z	$z = \dfrac{Z}{U^2{}_B} = \{Z\} \dfrac{100}{\{U^2{}_B\}} \dfrac{\%}{\text{MVA}}$	$'z = \dfrac{Z \cdot S_B}{U^2{}_B} = \{Z\} \dfrac{\{S_B\}}{\{U^2{}_B\}} \cdot 1$
Power S	$s = S = \{S\} \cdot 100\% \cdot \text{MVA}$	$'s = \dfrac{S}{S_B} = \dfrac{\{S\}}{\{S_B\}} \cdot 1$

Table 1.3   Conversions of quantities between %/MVA system and Ohm system

%/MVA system $\rightarrow$ Ohm system	Ohm system $\rightarrow$ %/MVA system
$\dfrac{U}{\text{kV}} = \dfrac{u}{\%} \cdot \dfrac{1}{100} \cdot \dfrac{U_B}{\text{kV}}$	$\dfrac{u}{\%} = \dfrac{U}{\text{kV}} \cdot 100 \cdot \dfrac{1}{U_B/\text{kV}}$
$\dfrac{I''_k}{\text{kA}} = \dfrac{i''_k}{\text{MVA}} \cdot \dfrac{1}{U_B/\text{kV}}$	$\dfrac{i''_k}{\text{MVA}} = \dfrac{I''_k}{\text{kA}} \cdot \dfrac{U_B}{\text{kV}}$
$\dfrac{Z}{\Omega} = \dfrac{z}{\%/\text{MVA}} \cdot \dfrac{1}{100} \left(\dfrac{U_B}{\text{kV}}\right)^2$	$\dfrac{z}{\%/\text{MVA}} = \dfrac{Z}{\Omega} \cdot 100 \cdot \dfrac{1}{(U_B/\text{kV})^2}$
$\dfrac{S''_k}{\text{MVA}} = \dfrac{s''_k}{\% \cdot \text{MVA}} \cdot \dfrac{1}{100}$	$\dfrac{s''_k}{\% \cdot \text{MVA}} = \dfrac{S''_k}{\text{MVA}} \cdot 100$

Table 1.4   Calculation of the impedances of electrical equipment in Ohms

Equipment	Impedance in positive phase-sequence system	Remarks
Synchronous machine (generator, motor, phase shifter)	$X_G = (x_d'' \cdot U_{rG}^2)/(100\% \cdot S_{rG})$ $R_{sG} = 0.05 \cdot X_G: S_{rG} \geq 100$ MVA $R_{sG} = 0.07 \cdot X_G: S_{rG} < 100$ MVA $R_{sG} = 0.12 \cdot X_G$	x_d'' saturated subtransient reactance in % S_{rG} rated apparent power for calculation of i_p for high voltage motors for calculation of i_p for low voltage motors
Transformer	$Z_T = (u_{kr} \cdot U_{rT}^2)/(100\% \cdot S_{rT})$ $R_T = (u_{Rr} \cdot U_{rT}^2)/(100\% \cdot S_{rT})$ $X_T = \sqrt{Z_T^2 - R_T^2}$	U_{rT} rated voltage, high voltage or low voltage side S_{rT} rated apparent power u_{kr} impedance voltage % for high voltage transformer, the following generally applies: $X_T \approx Z_T = (u_{kr} \cdot U_{rT}^2)/((100\%) \cdot S_{rT})$
Asynchronous motor	$X_M = (I_{rM}/I_a) \cdot (U_{rM}^2/S_{rM})$ $R_M = 0.1 \cdot X_M: \quad P_{rMp} \geq 1$ MW $R_M = 0.15 \cdot X_M: \quad P_{rMp} < 1$ MW $R_M = 0.42 \cdot X_M$	S_{rM} rated apparent power $S_{rM} = P_{rM}/(\eta \cdot \cos\varphi)$ I_a starting current I_{rM} rated current P_{rMp} rated active power high voltage motors low voltage motors including connecting cable
Current-limiting reactor	$X_D = (u_r \cdot U_{rD}^2)/(100\% \cdot S_{rD})$	S_{rD} rated apparent power $S_{rD} = \sqrt{3} \cdot U_{rD} \cdot I_{rD}$ I_{rD} rated current U_{rD} rated voltage u_r rated voltage drop
Impedance of system supply	$Z_Q = (1.1 \cdot U_{nQ}^2)/S_{kQ}''$ $X_Q = 0.995 \, Z_Q$ $R_Q = 0.1 \, X_Q$	S_{kQ}'' initial symmetrical short-circuit power at the system connection point Q U_{nQ} nominal voltage if precise values are not known
Overhead line or cable	$X_L = X_L' \cdot l$ $R_L = R_L' \cdot l$	X_L', R_L' in Ω/km in the circuit
Shunt reactor Shunt capacitor	$X_D = U_r^2/S_{rD}$ $X_C = U_r^2/S_{rC}$	S_{rD}, S_{rC} rated apparent power (three-phase) U_r rated voltage

Table 1.5   Calculation of the impedances of electrical equipment in %/MVA

Equipment	Impedance in positive phase-sequence system	Remarks
Synchronous machine (generator, motor, phase shifter) (G 3~) ///—	$x_G = x''_d / S_{rG}$ $r_{sG} = 0.05 \cdot X_G$: $S_{rG} \geq 100$ MVA $r_{sG} = 0.07 \cdot X_G$: $S_{rG} < 100$ MVA $r_{sG} = 0.12 \cdot X_G$	x''_d saturated subtransient reactance in % S_{rG} rated apparent power in MVA for calculation of i_p for high voltage motors for calculation of i_p for low voltage motors
Transformer —///—(○○)—///—	$z_T = u_{kr} / S_{rT}$ $r_T = u_{Rr} / S_{rT}$ $x_T = \sqrt{z^2_T - r^2_T}$	U_{rT} rated voltage, high voltage or low voltage side S_{rT} rated apparent power in MVA u_{kr} impedance voltage % for high voltage transformer, the following generally applies: $X_T \approx Z_T = u_{kr} / S_{rT}$
Asynchronous motor (M)—///—	$x_M = (I_{rM} / I_a) \cdot (100\% / S_{rM})$ $r_M = 0.1 \cdot x_M$: $P_{rMp} \geq 1$ MW $r_M = 0.15 \cdot x_M$: $P_{rMp} < 1$ MW $r_M = 0.42 \cdot x_M$	S_{rM} rated apparent power in MVA $S_{rM} = P_{rM} / (\eta \cdot \cos \varphi)$ I_a motorstarting current I_{rM} motor rated current P_{rMp} rated power high voltage motors low voltage motors including connecting cable
Current-limiting reactor —(⊃)—///—	$x_D = u_r \neq S_{rD}$	S_{rD} rated apparent power in MVA $S_{rD} = \sqrt{3} \cdot U_{rD} \cdot I_{rD}$ I_{rD} rated current U_{rD} rated voltage u_r rated voltage drop
Impedance of system supply [▨]—///—•Q—	$z_Q = 110\% / S''_{kQ}$ $x_Q = 0.995 \, z_Q$ $r_Q = 0.1 \, x_Q$	S''_{kQ} initial symmetrical short-circuit power at the system connection point Q in MVA U_{nQ} nominal supply voltage if precise values are not known
Overhead line or cable —[~]—///—	$x_L = (X'_L \cdot l \cdot 100\%) U^2_n$ $r_L = R'_L \cdot l \cdot 100\%) / U^2_n$	X'_L, R'_L in Ohm/km in the circuit U_n nominal voltage of supply system to which the line is located
Shunt reactor Shunt capacitor	$x_D = 100\% / S_{rD}$ $x_C = 100\% / S_{rC}$	$S_{rD}; S_{rC}$ rated apparent power (three-phase) in MVA U_r rated voltage

Table 1.4 gives an overview for calculation of the impedances of electrical equipment in Ohms and Table 1.5 for the calculation in %/MVA. A comparison of the two tables shows the great advantage of the %/MVA system, because the impedances can be calculated directly from the equipment characteristics (name plate data) and the calculation is easier than that of the Ohm system.

Unit, numerical value and magnitude equations are used for the calculation. In this case, unit equations are used, for instance between different systems, in order to convert the units (see Equation 1.52) for conversion of impedances from the %/MVA system to the Ohm system as follows:

$$1\,\Omega = 100/U_B^2 \cdot (\%/\text{MVA}) \qquad (1.52)$$

Numerical value equations are used for the fast calculation of quantities, whereby the particular output quantities may be used only in the defined units. Equation (1.53) is an example of the calculation of the initial symmetrical short-circuit current with a numerical value equation, as follows:

$$I''_{k3} = 110/(\sqrt{3} \cdot z_1)/U_n, \qquad (1.53)$$

whereby the initial symmetrical short-circuit current I''_{k3} in kA is calculated by using the short-circuit impedance z_1 in %/MVA and nominal voltage U_n in kV.

Magnitude equations are universally used where quantities with a numerical value and a unit (as given in Equation 1.54) for the calculation of the apparent power, are to be used, as follows:

$$\underline{S} = \underline{U} \cdot \underline{I}^* \qquad (1.54)$$

The result is a quantity, i.e. a numerical value with a unit.

Table 1.6 *Characteristics of overhead lines, values per km*
Depiction of a pylon arrangement without an earth wire.

Pylon shape	Conductor	U_n		Resistance	Reactance	Capacitance
		kV		Ω/km	Ω/km	nF/km
	50 Al	10 ... 20		0.579	0.355	8 ... 9
	50 Cu	10 ... 20		0.365	0.355	8 ... 9
	50 Cu	10 ... 20		0.365	0.423	8 ... 9
	70 Cu	10 ... 30		0.217	0.417	8 ... 9
	70 Al	10 ... 20		0.439	0.345	8 ... 9
	95 Al	20 ... 30		0.378	0.368	8 ... 9
	150/25	110		0.192	0.398	9

Table 1.7   Characteristics of transformers

U_{rOS}/U_{rUS}	S_r	u_{kr}	u_{Rr}
	MVA	%	%
MV/LV	0.05 ... 0.63	4	1 ... 2
	0.63 ... 2.5	6	1 ... 1.5
MV/MV	2.5 ... 25	6 ... 9	0.7 ... 1
HV/MV	25 ... 63	10 ... 16	0.6 ... 0.8

Low voltage: $U_n < 1$ kV
Medium voltage: $U_n = 1$ kV ... 66 kV
High voltage: $U_n > 66$ kV

Table 1.8   Characteristics for cables: resistances per km of a positive sequence system at 20°C in Ω/km

Conductor mm²	Resistance in Ω/km	
	Al	Cu
50	0.641	0.387
70	0.443	0.268
95	0.320	0.193
120	0.253	0.153
150	0.206	0.124
185	0.164	0.0991
240	0.125	0.0754
300	0.1	0.0601

1.5.4 Characteristics of typical equipment

To investigate the phenomena of system perturbations, it is often necessary to make rough calculations of the impedances of equipment. Because the topic of system perturbations is of interest, particularly in medium and low voltage systems, the characteristics of typical equipment such as transformers, overhead lines and cables from the aforementioned voltage levels are listed in the following table. In each case, however, the determining factor is the characteristics of the equipment used, which should be taken from name plates, data sheets or test records. Further examples for equipment data are given in the following references [4, 5, 6].

Table 1.9   Characteristics for paper-insulated cables: reactances per unit length of a positive sequence system in Ω/km

Conductor mm²	Reactance in Ω/km					
	A		B		C	
	1 kv	6 kV	10 kV	10 kV	20 kV	20 kV
50	0.088	0.1	0.1	0.11	0.13	0.14
70	0.085	0.1	0.1	0.1	0.12	0.13
95	0.085	0.093	0.1	0.1	0.11	0.12
120	0.085	0.091	0.1	0.097	0.11	0.12
150	0.082	0.088	0.092	0.094	0.1	0.11
185	0.082	0.087	0.09	0.091	0.1	0.11
240	0.082	0.085	0.089	0.088	0.097	0.1
300	0.082	0.083	0.086	0.085	0.094	0.1

A) Cables with steel band armouring
B) Three-core separately-sheathed cable
C) Single-core cables (triangular laying)

Table 1.10   Characteristics for cables: reactances per km of a positive or sequence system in Ω/km; using steel band armouring the reactances are increased by about 10%

Conductor mm²	Reactance in Ω/km					
	D			E		
	1 kV	6 kV	10 kV	1 kV	6 kV	10 kV
50	0.095	0.127	0.113	0.078	0.097	0.114
70	0.09	0.117	0.107	0.075	0.092	0.107
95	0.088	0.112	0.104	0.075	0.088	0.103
120	0.085	0.107	0.1	0.073	0.085	0.099
150	0.084	0.105	0.097	0.073	0.083	0.096
185	0.084	0.102	0.094	0.073	0.081	0.093
240	0.082	0.097	0.093	0.072	0.078	0.089
300	0.081	0.096	0.091	0.072	0.077	0.087

D) PVC multi-wire insulated cables
E) PVC single-core cables (triangular laying)

Table 1.11   Characteristics for cables: reactances per km of a positive sequence system in Ω/km; using steel band armouring the reactances are increased by about 10%

Conductor mm²	Reactance in Ω/km			
	F		G	
	1 kV	10 kV	1 kV	10 kV
50	0.072	0.11	0.088	0.127
70	0.072	0.103	0.085	0.119
95	0.069	0.099	0.082	0.114
120	0.069	0.095	0.082	0.109
150	0.069	0.092	0.082	0.106
185	0.069	0.09	0.082	0.102
240	0.069	0.087	0.079	0.098
300		0.084		0.095

F) XPET multi-wire insulated cables
G) XPET single-core insulated cables (triangular laying)

Table 1.12   Characteristics for paper-insulated cables: capacitance per km of a positive sequence system in μF/km

Conductor mm²	Capacitance in μF/km				
	A			B	
	1 kV	6 kV	10 kV	10 kV	20 kV
50	0.68	0.38	0.33	0.45	0.29
70	0.76	0.42	0.37	0.52	0.33
95	0.84	0.49	0.42	0.59	0.37
120	0.92	9.53	0.46	0.62	0.4
150	0.95	0.6	0.51	0.69	0.43
185	1.0	0.65	0.55	0.78	0.47
240	1.03	0.74	0.61	0.89	0.53
300	1.1	0.82	0.71	0.96	0.58

A) Belted insulating cable
B) Single-core cables and three-core separately-sheathed cables

Table 1.13 Characteristics for cables: capacitance per km of a positive sequence system in µF/km

Conductor	Capacitance in µF/km				
mm²	C			D	
	1 kV	6 kV	10 kV	10 kV	20 kV
50	k. A.	0.32	0.43	0.24	0.17
70	k. A.	0.35	0.48	0.28	0.19
95	k. A.	0.38	0.53	0.31	0.21
120	k. A.	0.43	0.58	0.33	0.23
150	k. A.	0.45	0.63	0.36	0.25
185	k. A.	0.5	0.7	0.39	0.27
240	k. A.	0.55	0.83	0.44	0.3
300	k. A.	0.6	0.92	0.48	0.32

C) PVC insulated cable
D) XPET insulated cable

1.6 Calculation examples

1.6.1 Graphical determination of symmetrical components

The corresponding voltages of the symmetrical components (012 system) are constructed for the voltage vectors $\underline{U}_R$, $\underline{U}_Y$ and $\underline{U}_B$ as shown in Figure 1.19.

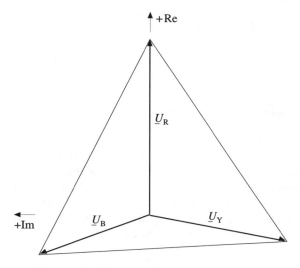

Figure 1.19 Vector diagram of voltages in RYB-system

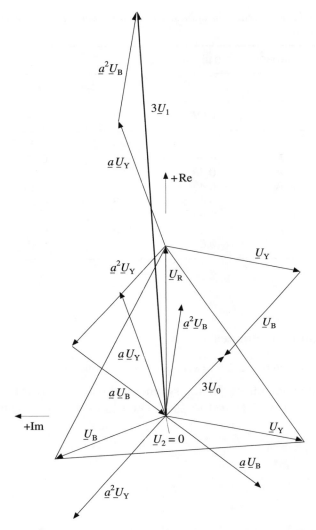

Figure 1.20   Construction of a vector diagram of symmetrical components based on Figure 1.19

The solution is shown in Figure 1.20. The resultant voltage of the negative sequence system is $\underline{U}_2 = 0$; with the voltages of the positive and negative sequence systems being $\underline{U}_1 \neq 0$ and $\underline{U}_0 \neq 0$ respectively. This is simply due to the fact that the three phase-to-earth voltages $\underline{U}_R$, $\underline{U}_Y$ and $\underline{U}_B$ are asymmetrical. The three phase-to-phase voltages are symmetrical. If the three phase-to-phase voltages are likewise asymmetrical then the voltage of the negative sequence system is $\underline{U}_2 \neq 0$.

1.6.2 *Arithmetical determination of symmetrical components*

The corresponding currents in the symmetrical components are calculated for the currents of the RYB system:

$$\underline{I}_R = 0 \text{ kA}; \underline{I}_S = 1 \text{ kA} + \text{j}5 \text{ kA}; \underline{I}_T = -1 \text{ kA} + \text{j}5 \text{ kA}$$

Solution: Conversion into polar form gives us:

$$\underline{I}_R = 0 \text{ kA}e^{\text{j}0}; \underline{I}_S = 5.01 \text{ kA } e^{\text{j}78.69}; \underline{I}_T = 5.01 \text{ kA } e^{\text{j}101.31}.$$

Application of Equation (1.20) for the current gives us:

$$\underline{I}_0 = (1/3)(\underline{I}_R + \underline{I}_S + \underline{I}_T)$$
$$= (1/3)(0 \ e^{\text{j}0} + 5.01 \ e^{\text{j}78.69} + 5.01 \ e^{\text{j}101.31}) \text{ kA}$$
$$\underline{I}_1 = (1/3)(\underline{I}_R + \underline{a}\underline{I}_S + \underline{a}^2\underline{I}_T)$$
$$= (1/3)(0 \ e^{\text{j}0} + 5.01 \ e^{\text{j}78.69} \cdot e^{\text{j}120} + 5.01 \ e^{\text{j}101.31} \cdot e^{\text{j}240}) \text{ kA}$$
$$= (1/3)(0 \ e^{\text{j}0} + 5.01 \ e^{\text{j}98.69} + 5.01 \ e^{\text{j}341.31}) \text{ kA}$$
$$\underline{I}_2 = (1/3)(\underline{I}_R + \underline{a}^2\underline{I}_S + \underline{a}\underline{I}_T)$$
$$= (1/3)(0 \ e^{\text{j}0} + 5.01 \ e^{\text{j}78.69} \cdot e^{\text{j}240} + 5.01 \ e^{\text{j}101.31} \cdot e^{\text{j}120}) \text{ kA}$$
$$= (1/3)(0 \ e^{\text{j}0} + 5.01 \ e^{\text{j}318.69} + 5.01 \ e^{\text{j}221.31}) \text{ kA}$$

By resolution we get the following:

$$\underline{I}_0 = \text{j}3.275 \text{ kA}$$
$$\underline{I}_1 = -\text{j}1.070 \text{ kA}$$
$$\underline{I}_2 = \text{j}2.204 \text{ kA}$$

Taking rounding errors into consideration, it is apparent that the sum of the currents of the symmetrical components is, in this case, zero. The current conditions then apply for a two-phase short-circuit with earth.

1.6.3 *Calculation of equipment*

Calculations for the reactances and resistances are made for the following equipment in the %/MVA system and the Ω system.
Synchronous machine:

$$S_{rG} = 50 \text{ MVA}; U_{rG} = 10.5 \text{ kV}; \cos \varphi_{rG} = 0.8; x''_d = 14.5\%$$

Two-winding transformer:

$$S_{rT} = 50 \text{ MVA}; U_{rTOS}/U_{rTUS} = 110 \text{ kV}/10.5 \text{ kV}; u_{kr} = 10\%$$
$$u_{Rr} = 0.5\% \text{ or } P_{Vk} = 249 \text{ kW}$$

System, at the system connection point Q:

$$S''_{kQ} = 2\ 000 \text{ MVA}; U_{nQ} = 110 \text{ kV}$$

Three-phase XPET cable (N2XSY 18/30 kV 1 × 500 RM/35):

$$R'_L = 0.0366 \ \Omega/\text{km}; X'_L = 0.112 \ \Omega/\text{km}; l = 10 \text{ km}; U_n = 30 \text{ kV}$$

Short-circuit limiting reactor:

$$u_{rD} = 5\%; \; I_{rD} = 500 \text{ A}; \; U_n = 10 \text{ kv}$$

The calculations for the %MVA system or Ω system can be performed using the conversion equations listed in Table 1.3.

Solution: Synchronous machine in the %/MVA system:

$$x_G = x''_d/S_{rG} = 14.5\%/50 \text{ MVA} = 0.29\%/\text{MVA}$$
$$r_G = 0.07 \; x_G = 0.0203\%/\text{MVA, since } S_{rG} < 100 \text{ MVA and } U_{rG} < 1 \text{ kV}$$

Synchronous machine in the Ω system:

$$X_G = (x''_d \cdot U_{rG}^2)/(S_{rG} \cdot 100\%) = (14.5\% \cdot (10.5 \text{ kV})^2)/(50 \text{ MVA} \cdot 100\%)$$
$$= 0{,}2304 \; \Omega$$
$$R_G = 0.07 \cdot X_G = 0.0161 \; \Omega$$

Two-winding transformer in the %/MVA system:

$$z_T = u_{kr}/S_{rT} = 10\%/50 \text{ MVA} = 0.2\%/\text{MVA}$$
$$r_T = u_{Rr}/S_{rT} = 0{,}5\%/50 \text{ MVA} = 0.01\%/\text{MVA}$$
$$x_T = \sqrt{z_T^2 - r_T^2} = 0.1997 \; \% \neq \text{MVA}$$

Two-winding transformer in the Ω system:

$$Z_T = (u_{kr} \cdot U_{rTOS}^2)/(S_{rT} \cdot 100\%) = (10\% \cdot (110 \text{ kV})^2)/(50 \text{ MVA} \cdot 100\%)$$
$$= 24.2 \; \Omega, \text{ related to } 110 \text{ kV}$$
$$R_T = (u_{Rr} \cdot U_{rTOS}^2)/(S_{rT} \cdot 100\%) = (0.5\% \cdot (110 \text{ kV})^2)/(50 \text{ MVA} \cdot 100\%)$$
$$= 1{,}21 \; \Omega, \text{ related to } 110 \text{ kV}$$
$$X_T = \sqrt{Z_T^2 - R_T^2} = 24.17 \; \Omega, \text{ related to } 110 \text{ kV}$$

Supply, at connection point Q in the %/MVA system:

$$z_Q = 110\%/S''_{kQ} = 110\% \neq 2\,000 \text{ MVA} = 0.055\%/\text{MVA}$$
$$x_Q = 0.995 \cdot z_Q = 0.0547\%/\text{MVA}$$
$$r_Q = 0.1 \cdot z_Q = 0.0055\%/\text{MVA}$$

Supply, at connection point Q in the Ω system:

$$Z_Q = 1.1 \cdot U_{nQ}^2/(S''_{kQ} \cdot 100\%) = 1.1 \cdot (110 \text{ kV})^2/(2\,000 \text{ MVA} \cdot 100\%)$$
$$= 0.06655 \; \Omega$$
$$X_Q = 0.995 \cdot Z_Q = 0.06622 \; \Omega$$
$$R_Q = 0.1 \cdot z_Q = 0.0066 \; \Omega$$

Three-phase cable (N2XSY 18/30 kV 1 × 500 RM/35) in the %/MVA system:

$$r_L = (R'_L \cdot l \cdot 100\%)/U_n^2 = (0.0366 \; \Omega/\text{km} \cdot 10\text{km} \cdot 100\%)/(30 \text{ kV})^2$$
$$= 0.041\%/\text{MVA}$$
$$x_L = (X'_L \cdot l \cdot 100\%)/U_n^2 = (0.112 \; \Omega/\text{km} \cdot 10\text{km} \cdot 100\%)/(30 \text{ kV})^2$$
$$= 0.124\%/\text{MVA}$$

Three-phase cable (N2XSY 18/30 kV 1 × 500 RM/35) in the Ω system:

$$R_\text{L} = R'_\text{L} \cdot l = 0.0366 \ \Omega/\text{km} \cdot 10\text{km} = 0.366 \ \Omega$$
$$X_\text{L} = X'_\text{L} \cdot l = 1.112 \ \Omega/\text{km} \cdot 10\text{km} = 0.12 \ \Omega$$

Short-circuit limiting reactor in the %/MVA system:

$$x_\text{D} = u_\text{rD}/(\sqrt{3} \cdot U_\text{rD} \cdot I_\text{rD}) = 5\%/(\sqrt{3} \cdot 10 \ \text{kV} \cdot 0.5 \ \text{kA}) = 0.577\%/\text{MVA}$$

Short-circuit limiting reactor in the Ω system:

$$X_\text{D} = (u_\text{rD} \cdot U_\text{rD}^2)/(\sqrt{3} \cdot U_\text{rD} \cdot I_\text{rD} \cdot 100\%)$$
$$= (5\% \cdot (10\text{kV})^2)/(\sqrt{3} \cdot 10 \ \text{kV} \cdot 0.5 \ \text{kA} \cdot 100\%) = 0.577 \ \Omega$$

The impedance in the %/MVA system and in the Ω system has the same numerical value as the reference voltage is 10 kV.

1.7 References

1   'Deutschland und die Welt (Germany and the world)', *Frankfurter Allgemeine Zeitung* 28. 2. 97
2   SCHLABBACH, J. 'Elektroenergieversorgung—Betriebsmittel und Auswirkungen der elektrischen Energieverteilung (Electrical energy supply sytems—components and effects of electrical energy distribution)' (VDE-VERLAG: Berlin and Offenbach, 1995)
3   HOSEMANN, G., and BOECK, W.: 'Grundlagen der elektrischen Energietechnik (fundamental principles of electrical energy technology)', (Springer-Verlag: Berlin, Heidelberg, New York, 1979)
4   BOSSE, G.: 'Grundlagen der Elekrotechnik I–IV (fundamental principles of electrical engineering I–IV)' (Bibliographisches Institut, Mannheim, 1973)
5   WEßNIGK, K.-D.: 'Kraftwerkselektrotechnik (power station electrotechnology)' (VDE-VERLAG: Berlin and Offenbach, 1993)
6   ABB: 'Switchgear manual'. (Cornelsen-Verlag, 1987, 8th edn.)

Chapter 2

Harmonics and interharmonics

2.1 Occurrence and causes

2.1.1 General

Harmonics occur due to equipment with non-linear characteristics such as transformers and fluorescent lamps, and today are principally due to power electronics components such as rectifiers, triacs or thyristors. In this regard, particular attention should be paid to the use of rectifiers with capacitor smoothing which are used extensively in televisions, PCs and compact fluorescent lamps, especially in domestic and office environments. Based on research by the VDEW (German Association of Electric Utilities) in the early 1990s, 25% of domestic loads were attributable to electronic loads; i.e. lighting 3%, consumer electronic equipment 21% and controlled drives (washing machines) 1%. If we further consider that the proportion of domestic loading is 27% of the total system load then, for Germany in 1992, domestic electronic loading amounted to 6.7% of the total system loading or some 4 GW. This trend is increasing.

2.1.2 Occurrence due to network equipment

Only multiples of the fundamental frequency occur in equipment with non-linear characteristics such as transformers and discharge lamps. As a first example, the non-linear $H(B)$ characteristic of a transformer as seen in Figure 2.1 is explained. The hysteresis is disregarded in this case.

From a pure sinusoidal supply voltage (without harmonics)

$$u(t) = \sqrt{2}\ U \cos(\omega t + \varphi_u) \tag{2.1}$$

we obtain from the magnetic flux

$$\Phi_\mu = \int u\ \mathrm{d}t \tag{2.2}$$

the magnetic flux density B in steady-state condition:

$$B = d\Phi/dA \tag{2.3a}$$

$$b(t) = \sqrt{2}\, B \sin{(\omega t + \varphi_u)} \tag{2.3b}$$

By application of Ampere's law for the $H(B)$ characteristics of the transformer we get

$$\oint_s H\,ds = \int_A J\,dA \tag{2.4a}$$

$$\int_A J\,dA = \Theta = N\,I \tag{2.4b}$$

the harmonics-impressed current course

$$i(t) = \sum_{h=1}^{\infty} \sqrt{2}I_h \sin{(h\,\omega_1\,t + \varphi_{1h})} \tag{2.5}$$

The $H(B)$ characteristic in Figure 2.1 is described using a polynomial of the n-th

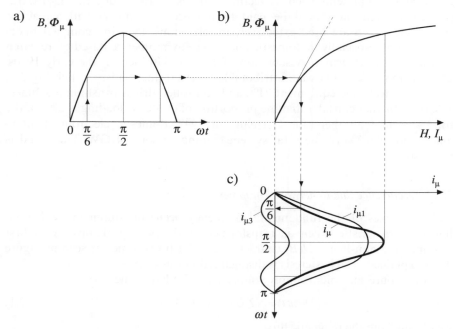

Figure 2.1   *Graphical determination of the magnetising current of a transformer*
    a) time course of voltage and magnetic induction
    b) H(B)-characteristic of transformer with iron core (not to scale)
    c) time course of magnetising current; basic frequency and 3rd harmonic

order; n is odd due to the central symmetry (see also section 1.4.2). Therefore, harmonics of an odd order h are produced for the current $i(t)$.

As a second example, the occurrence of harmonics by three-phase generators is discussed. The assumption that the voltages produced by three-phase generators are truly sinusoidal is essentially incorrect, since this presupposes that the individual turns of the stator winding of a synchronous generator are distributed evenly over the circumference and are not laid into discrete slots.

By first considering the course of the induced voltage of a single winding ($q = 1$) as shown in Figure 2.2a, it can be seen that it is rectangular in shape, as is the course of the magnetomotive force. Fourier analysis produces solely odd-numbered harmonics for the curve trace. By increasing the number of windings q, a stepped curve is obtained for the magnetomotive force or induced voltage by summation of the corresponding magnetomotive forces of the individual windings as in Figure 2.2b, which are one slot pitch out of phase relative to each other. The courses of the magnetomotive force and the induced voltage thus approach that of the ideal sinusoidal form, the amplitudes of the harmonics are reduced and, in some cases, harmonics are eliminated.

At present, non-sinusoidal voltages are also generated in electrical energy

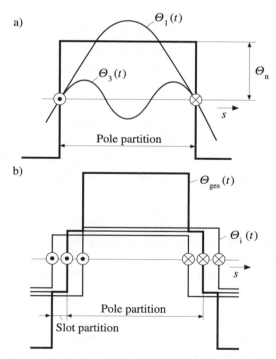

Figure 2.2   *Part of a cross section of an electrical machine (not to scale) with magnetic induction*
a) single winding (q = 1); basic frequency and 3rd harmonic
b) three windings (q = 3)

generating systems using regenerative energy sources. This is due to the fact that, for example, photovoltaic systems are connected to three-phase a.c.-systems using inverters. In the field of wind energy converters, variable-speed coupling using power electronics components is also widespread. Here, the differences between supply equipment and consumers with regard to the origin of supply perturbations are slight due to the use of power electronics. The occurrence of harmonics and interharmonics due to power electronics is explained in the following section.

2.1.3 Occurrence due to power electronics equipment

2.1.3.1 Basic principles

As explained in section 1.4.2, each period T or 2π periodic time function can be shown as a superimposition of sinus- and cosinus-functions (Fourier synthesis).

The amplitude spectrum of, for example, the voltage during timed switching of consumers using phase control (thyristor controller) or multicycle control (thermal equipment) can be mathematically described by multiplication of the original function

$$u(t) = \sqrt{2}U \sin(\omega_1 t) \tag{2.6}$$

with the rectangular switching function S(t), whose spectrum is a function of $h\omega_s$, where $h = 0, 1, 2 \ldots$. The original function $u(t)$ contains voltage harmonics according to

$$u(t) = \sum_{h=1}^{n} \sqrt{2}U_h \sin(h\omega_1 t), \tag{2.7}$$

by multiplication with the rectangular switching function S(t) the voltage spectrum occurs at the consumer with a frequency according to

$$\omega_h = h\omega_1 \pm h\frac{\omega_s}{2\pi} \tag{2.8}$$

In the case of symmetrical phase-angle control with an almost constant control angle, as shown in Figure 2.3, $\omega_s = 2\omega_1$ applies. The voltage present at the consumer contains voltage harmonics of an order according to Equation (2.9) where $h = 1, 2, 3 \ldots$

$$\omega_h = h\omega_1 \tag{2.9}$$

Interharmonics occur in case of periodically variable control angle strategies.

As an example, multicycle control is performed using multiples of the system frequency as the control frequency. Here is the periodic duration of the switching frequency

$$T_S = mT \tag{2.10}$$

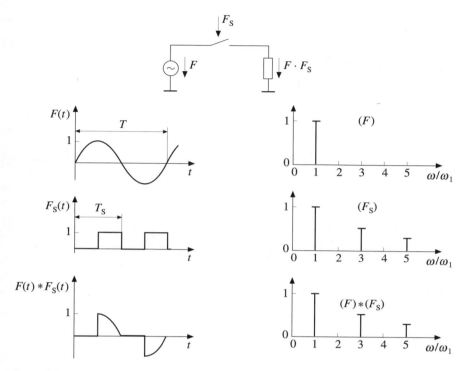

Figure 2.3　Time course and harmonics of a non-sinusoidal voltage; synchronised switching

where $m = 2, 3, \ldots$, and since

$$\omega = \left(\mu \pm \frac{h}{m}\right)\omega_1 \tag{2.11}$$

mainly interharmonics occur. Figure 2.4 clarifies the relationships.

2.1.3.2 Full-wave rectifier with capacitor smoothing

The full-wave rectifier with capacitor smoothing shown in Figure 2.5 is first considered as an example of the widespread use of power electronics. Starting from the steady state at $t = 0$, the system voltage $u(t)$ increases. The voltage on the d.c. side falls, relative to the time-constants of the connected load, consisting of smoothing capacitor C and load R. If the supply voltage becomes higher than the voltage on the d.c. side (irrespective of the conducting voltage of the diodes) current will flow, thus charging the capacitor. As the voltage is now lower due to re-charging of the capacitor, the higher voltages on the d.c. voltage side bias-off the diodes, and the charging current ceases. This process is repeated for the negative half-wave of the supply voltage. Time point, time duration and level of charge-current pulses are, in this case, dependent on the values of the smoothing capacitor and the series impedance.

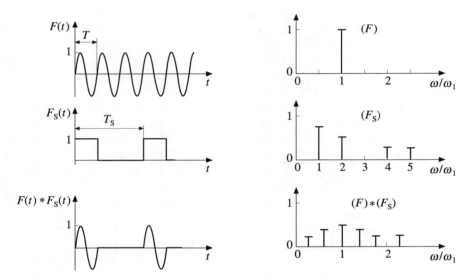

Figure 2.4   Time course, harmonics and interharmonics of a non-sinusoidal voltage; pulsed switching

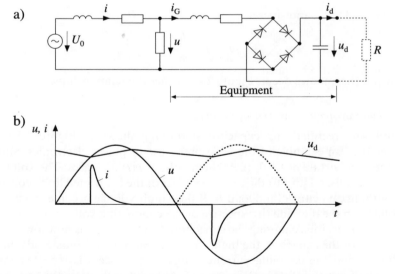

Figure 2.5   a.c./d.c.-converter with capacitor smoothing (Graetz-bridge)
a) electrical diagram with system connection
b) time course of current and voltages

The smoothing capacitor is typically recharged about 2 ms prior to maximum voltage being reached. Since all devices operated in the equipment incorporating full-wave rectifiers, for example consumer electronic equipment, compact fluorescent lamps or primary timed switched-mode power supplies behave in a similar fashion and have roughly the same phase position for their respective order of harmonic currents, this results in a pulsed loading of the supply and a high harmonic loading. This characteristic is expressed using the co-phasal factor, which is defined as a quotient of the geometrical sum to the arithmetical sum of the relevant harmonic currents of the same order h of the various consumers.

Research [1] indicated that by using a number of compact fluorescent lamps, whose behaviour is similar to a full-wave rectifier with capacitor smoothing, a considerable reduction in co-phasal factor is obtained. The reason for this is that compact fluorescent lamps are relatively insensitive to the ripple of d.c. voltage and are therefore extremely flexible with regard to the recharging time point and duration, while the power supplies of consumer electronics require a more consistent d.c. voltage and are therefore less flexible with regard to recharge time point and duration. The operation of several power supplies with capacitor smoothing therefore scarcely reduces the phase angle.

The measured current course and respective frequency spectrum of the current of a full-wave rectifier with capacitor smoothing (primary timed switched-mode power supply) is shown in Figure 2.6. The even-numbered harmonics can be disregarded. However, the odd-numbered harmonics up to the higher orders contain a significant component. The components of harmonics of orders $h = 3$, 5, 7, relative to the fundamental component of current are, $I_3/I_1 = 85\%$; $I_5/I_1 = 80\%$; $I_7/I_1 = 60\%$.

2.1.3.3 Three-phase bridge circuit

Three-phase bridge circuits are now in widespread use in industrial applications, in the form of uncontrolled, controlled, six-, twelve- and higher- pulse circuits. The basic configuration of a six-pulse, controlled, three-phase bridge circuit is shown in Figure 2.7, with which the behaviour of this harmonics generator can be clarified.

By applying a firing pulse and a positive anode-cathode-voltage to the thyristor it can be brought into a conductive state. A current thus flows from the three-phase side of the circuit to the load side. By synchronising the firing pulses, a thyristor in both the positive and negative bridge-halves can be brought into a conductive state. The sequence of the firing pulses is therefore determined by the phase sequence of the input voltage.

When a thyristor is fired, the current commutates from the active thyristor to the newly fired thryistor of the same section of the bridge. Because of the time course of the phase voltage of the three-phase system, a negative anode-cathode-voltage occurs at the current-conducting thyristor and the current in this thyristor ceases to flow. The commutation time is finite, during which both of the thyristors subjected to the commutation process are active. The

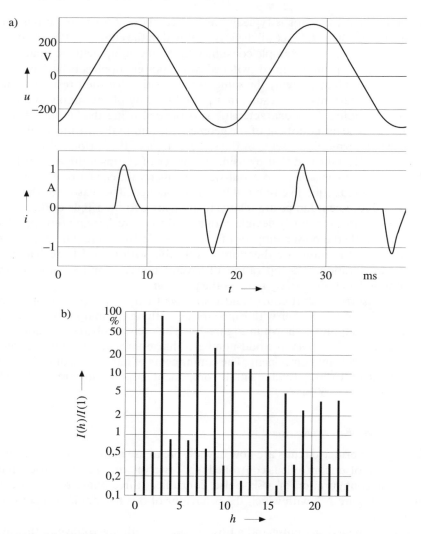

Figure 2.6   Measured time course and harmonics of a.c.-current of a.c./d.c.-converter with
capacitor smoothing [5]
a)   measured time course
b)   calculated harmonics

commutation time is dependent on the actual reactances present in the three-phase side.

Synchronous through-switching with power system frequency of the three-phase system from the supply side to the load side produces a d.c. voltage, the mean value of which can be varied by altering the gate-controlled turn-on time of the thyristors. If the thyristors are fired at the earliest possible moment, the

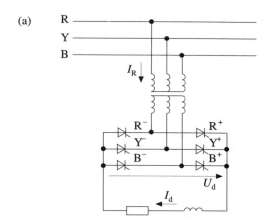

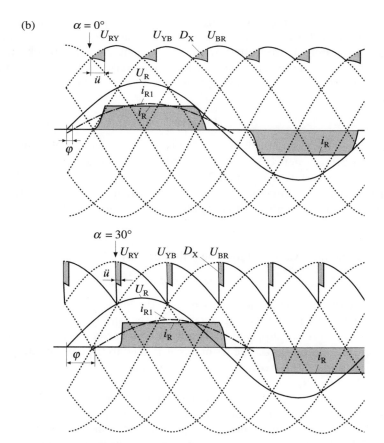

Figure 2.7   Six-pulse three-phase thyristor bridge
    a) electrical diagram
    b) time course of currents and voltages on a.c. side

mean value of the d.c. voltage will be maximal and the value reached will be the so-called ideal d.c. voltage in accordance with

$$U_{di} = \frac{3\sqrt{2}}{\pi} U \tag{2.12}$$

U is the r.m.s. value of the phase-to-earth voltage of the three-phase side.

If the gate-controlled turn-on time of the thyristor is retarded relative to the earliest possible moment, i.e. the natural commutation time point, then the mean value of the d.c. voltage is reduced in accordance with

$$U_{d\alpha} = \frac{3\sqrt{2}}{\pi} U \cos \alpha \tag{2.13}$$

α is the control angle (control angle $\alpha = 0°$ means the natural gate-controlled turn-on time).

If we assume that the load in the d.c. circuit is purely inductive, then a current for each $T/3$ (120°) will flow in the thyristor branches; the individual blocks of each branch are shifted by $T/3$ (120°). These current blocks also flow in the input lines on the supply side of the three-phase bridge. Fourier analysis of the current course on the three-phase side produces harmonics in accordance with

$$h = np \pm 1 \tag{2.14}$$

where $n = 1, 2, 3, \ldots$ and the number of pulses p is the number of commutations occurring during a network period. A three-phase bridge circuit with a pulse number of $p = 6$ is shown in Figure 2.7.

Harmonics of the orders $h = 5, 7, 11, 13, \ldots$ are generated accordingly. The level of harmonic currents is roughly in line with Equation (2.15a):

$$I_h/I_1 \approx 1/h \tag{2.15a}$$

where I_1 is the r.m.s. value at fundamental frequency

$$I_1/I_d = \sqrt{6}/\pi \tag{2.15b}$$

and I_d is the r.m.s. value of the d.c. current on the load side.

The commutating reactances (reactances of the supply system) limit the rise of line currents on the supply side. The harmonic currents shown in Equation (2.15a) are thereby reduced. The reduction is related to the commutating time or overlapping angle $ü$. The reduction factors r_h related to the overlapping angle are shown in Figure 2.8.

The ripple of the d.c. current leads to a further reduction in the harmonics of orders $h > 5$ in the six-pulse, three-phase bridge circuit. The reduction in harmonics is related to control angle α. The fifth harmonic is exceptional as the ripple of the d.c. current leads to a harmonic which is clearly higher in comparison with Equation (2.15a). The control angle too has almost no influence on the level of the fifth harmonic.

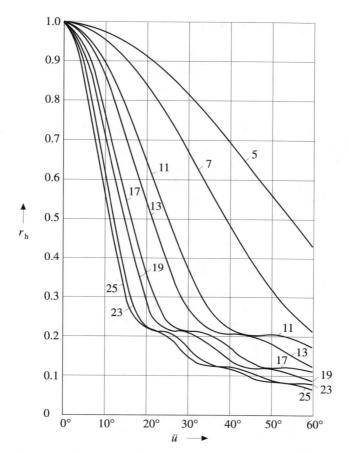

Figure 2.8 Factor r_h of harmonic RMS of six-pulse three-phase thyristor bridge related to overlapping angle *ü; ideally smoothed d.c. current*

If two, six-pulse, three-phase bridge circuits are connected to a supply system via two transformers with differing vector groups, e.g. Yy0 and Yd5, as seen in Figure 2.9, each three-phase bridge circuit will generate harmonics in accordance with Equation (2.14).

Because of the differing vector groups of the transformers, harmonics of the orders $h = 5, 7, 17, 19$ etc. with a phase shift of 180° are transmitted from the two transformers to the power supply side.

These harmonic components are completely quenched, provided that:

– the transformation ratios and the impedance voltages (commutating reactances) of the transformers are equal,
– all components are symmetrically configured,
– each of the three-phase bridges are operated using the same control angle,

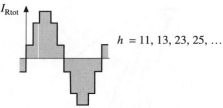

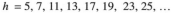

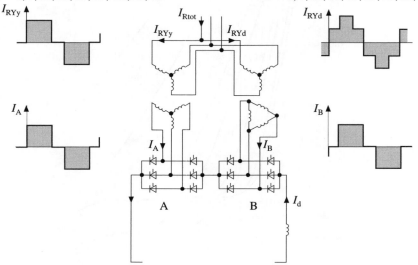

Figure 2.9   Electrical diagram of twelve-pulse three-phase thyristor bridge (series connection) with idealised time course of currents

–   the d.c. intermediate circuits have the same level of ripple,
–   the firing pulses are synchronised and
–   the fundamental frequency currents of each three-phase bridge are equal.

The complete circuit produces solely harmonics of the orders $h = 11, 13, 23, 25,$ ..., in effect acting as a twelve-pulse, three-phase bridge circuit in the supply.

The current harmonics on the supply side of a transformer of vector group Yy0 (or Dd0) for example, can for conductor R, be calculated as:

$$I_{RYy} = \frac{2\sqrt{3}}{\pi} I_d \left(\sin \omega t - \frac{1}{5} \sin 5\omega t - \frac{1}{7} \sin 7\omega t + \right.$$

$$\left. + \frac{1}{11} \sin 11\omega t + \frac{1}{13} \sin 13\omega t - \ldots \right) \qquad (2.16a)$$

The current for conductor R for a transformer of vector group Yd5 (or Dy5) is given as follows:

$$I_{RYd} = \frac{2\sqrt{3}}{\pi} I_d \left(\sin \omega t + \frac{1}{5} \sin 5\omega t + \frac{1}{7} \sin 7\omega t + \right.$$

$$\left. + \frac{1}{11} \sin 11\omega t + \frac{1}{13} \sin 13\omega t + \ldots \right) \tag{2.16b}$$

Therefore, the total current for a twelve-pulse converter can be calculated as:

$$I_{Rges} = I_{RYy} + I_{RYd} \tag{2.17a}$$

$$I_{Rges} = \frac{4\sqrt{3}}{\pi} I_d \left(\frac{1}{11} \sin 11\omega t + \frac{1}{13} \sin 13\,\omega t + \ldots \right) \tag{2.17b}$$

Twelve-pulse or higher converter circuits can also be achieved by other means, which are not dealt with further in this connection. Comprehensive descriptions are given in [2, 3, 4] and in other texts.

Where there are three-phase bridge connections, so-called non-characteristic harmonics occur which go beyond the ideal harmonic spectrum dealt with in the introduction, which are caused by:

– dynamic processes,
– timed firing pulses which are not precisely synchronised,
– asymmetries of supply system feed,
– asymmetries of components, particularly of the converter transformer in the case of higher-pulse converter circuits.

The level of this harmonic content normally remains within the range of a few percentage points relative to the fundamental component.

2.1.3.4 Converters

If instead of the three-phase bridge circuits with a d.c. current circuit or d.c. voltage circuit on the load side, as described in section 2.1.3.3, power electronics are used to convert the supply-side 50 Hz three-phase system to a three-phase system of variable frequency, variable voltage and with any number of phase, these are generally referred to as converters. The following are some examples:

– direct converters without an intermediate circuit,
– intermediate circuit converters with impressed voltage or impressed current in the intermediate circuit and a self-commutated inverter on the load side,
– intermediate circuit converters with impressed current in the intermediate circuit and load-commutated inverters on the load side, for supply of a converter motor,
– subsynchronous converter cascade.

The principle can be explained using the direct converter shown in Figure 2.10 as an example. At the supply side, the converter is connected with the six-pulse or higher three-phase bridge circuit already mentioned, which produces the harmonic spectrum shown in Equation (2.14) and explained in section 2.1.3.3.

$$h = np \pm 1 \tag{2.14}$$

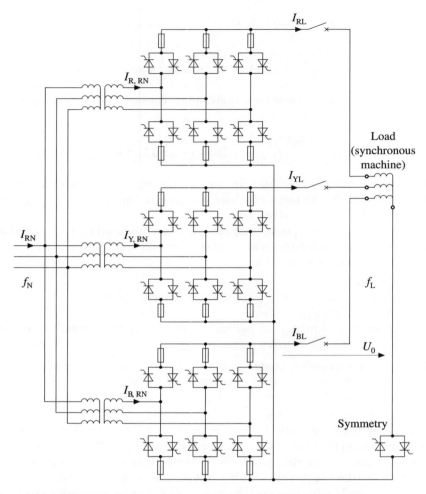

Figure 2.10   Electrical diagram of a frequency converter (direct converter)

where $n = 0, 1, 2, 3, \ldots$. In addition, currents with frequencies of f_j also occur, which are linked to frequency f_L of the output voltage (load) as follows:

$$f_j = 2\,m\,q\,f_L \qquad (2.18)$$

where $m = 0,1,2,3, \ldots$ and q is the number of conductors of the a.c. system of the load (number of windings of the connected motor). Because the frequency of the system at the load side is variable, under certain circumstances the current components with these frequencies are not whole-number multiples of the power supply frequency and are therefore designated interharmonic. The r.m.s. values of the interharmonics are generally less than 5% relative to the fundamental component of current, and decrease with increasing frequency.

We then get the following frequency spectrum for the direct converter.

$$f_i = (np \pm 1) f_N \pm 2\, m\, q\, f_L \tag{2.19}$$

where $m, n = 0, 1, 2, 3, \ldots$.

Frequency components with negative characteristics are important because their associated rotary field forms a contrarotating torque, and the associated currents and voltages represent a negative sequence system.

The frequency spectrum for the following output values is defined below for the converter shown in Figure 2.10:

Supply converter: number of pulses $p = 6$; $f_N = 50$ Hz

Load converter: number of windings of machine $q = 3$; $f_L = 7.14$ Hz

The following frequencies (harmonics) $f_i = (n\, p \pm 1)\, f_N$ are generated in the supply converter, i.e. in addition to the fundamental frequency of 50 Hz, the frequencies of 250 Hz, 350 Hz, 550 Hz etc. are also generated. The frequencies occurring in the supply current and load converter amount to $f_j = \pm 2\, m\, q\, f_L$. This results in the frequencies listed in Table 2.1.

If a pulse converter with a clock frequency in the 100 Hz–kHz range is employed as the rectifier on the load side, only high-frequency currents will occur, due to the converter on the load side. In general, these will not be transmitted to the supply side of the converter. Such converters behave as six-pulse or higher, three-phase bridge circuits in the supply, depending on the number of pulses on the three-phase side.

Converter circuits with respect to the generation of harmonics and interharmonics are dealt with in more detail in [4].

Table 2.1 Example of frequency components in the current of a direct converter, given in Hertz

Frequency component, produced by:		
Supply converter	± Load converter	= Sum
50	0.0	50.0
	42.84	7.16/92.84
	85.68	−35.68/135.68
	128.52	−78.52/178.52
250	0.0	250.0
	42.84	207.16/292.84
	85.68	164.32/335.68
	128.52	121.48/378.52
350	0.0	350.0
	42.84	307.16/392.84
	85.68	264.32/435.68
	128.52	221.48/578.52

Number of pulses of supply converter $p = 6$; $f_N = 50$ Hz;
Number of windings of machine (load converter) $q = 3$; $f_L = 7.14$ Hz

2.1.4 Occurrence due to random consumer behaviours

Powerful harmonics generators, principally used in industrial environments, are very predictable in their operation. By contrast, the operation of small consumers in domestic and light industrial environments can only be described using random averaging due to differing user habits.

The main causes of harmonics in domestic, light industrial and industrial environments are shown in Table 2.2.

Equipment with power ranges of 100 W up to several kW used in domestic and light industrial environments are, with few exceptions, designed to run on low voltage, a.c. current (single-phase) supplies, while equipment for use in industrial environments generally employs three-phase current. The most common converter circuit found in low voltage supplies is the full-wave rectifier with capacitor smoothing, also known as the peak-value rectifier. This circuit can be economically manufactured and is firmly established in the consumer electronics market. Rectifiers of this type are used in television sets, video recorders, satellite receivers, stereo systems, lighting, personal computers, accumulator chargers and, increasingly, also in high-power devices such as washing machines and air conditioning units. Compact fluorescent lamps with electronic series components behave in a manner similar to full-wave rectifiers in the supply.

Table 2.2   List of typical harmonics generators in domestic, light industrial and industrial use [6, 14]

Domestic and light industry	Industry and power supply companies
Converters	
Audio and video equipment	Induction furnaces
Halogen lamps	Traction supply converters
Compact fluorescent lamps	D.C. telecommunication networks
Dimmers	Regulated three-phase drives
Mixers and cutters	D.C. drives
Fridges and freezers	Machine tools
Microwave ovens	Welding equipment
Vacuum cleaners	Wind energy converters
Washing machines	Photovoltaic systems
Dishwashers	HVDC systems
Computers	
Pumps	
Non-linear U/I-characteristics	
Fluorescent lamps without electronic series control gear	Arc furnaces
	Arc welding equipment
Incandescent bulbs	Gas-discharge lamps
Small motors	Transformers
	Induction furnaces

Harmonic voltages in public medium and low voltage supplies, in particular fifth-order harmonic voltages, can, to a considerable degree, be linked to the use of full-wave rectifiers with capacitor smoothing. Several of the disadvantages in relation to the emission of harmonic currents should be noted:

- widespread propagation due to use in almost all consumer electronic equipment,
- high simultaneity of use (television sets, lighting) in particular in evenings and at weekends,
- high relative harmonic currents (see Figure 2.6),
- high level of harmonic currents of the same phase angle from various equipment (exception: compact fluorescent lamps).

Figure 2.11 shows the course of the third, fifth and seventh harmonic voltages in a 10-kV system (residential area with light industry) for autumn 1995. The course of the fifth harmonic shows a typical, relatively flat, almost constant path throughout the day, with a very distinct increase during the evening hours, the peak being reached between 20.00 and 21.00 hours. Due to the distribution of small consumers, none of which contribute in particular to the loading of harmonic components, the course of the fifth harmonic is steady, without any jumps in the nominal values. The maximum values are attributable to television sets as the main cause of the harmonic peaks in the evening.

However, the course of the third harmonic voltage seen in Figure 2.11 is essentially constant. Although currents of the third harmonic are undoubtedly

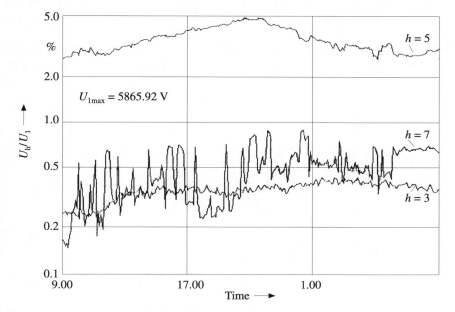

Figure 2.11   Time course of selected harmonics (logarithmic scale) in a 10 kV system Saturday, 4/11/95; system load P = 22.3 MW; residential area

generated by the full-wave rectifiers, only a small component reaches the primary supply due to the extremely high, zero-sequence impedance of the 10/0.4 kV transformers (Dy or Dz). In contrast, the course of the seventh harmonic voltage shows a periodicity of about one hour. This is clearly due to the effects of an industrial consumer which is also the underlying cause of the random course of the fifth harmonic voltage.

The causes are more clearly seen by measuring the fifth harmonic voltage over the course of an entire week. As an example, Figure 2.12 shows the course of the fifth harmonic voltage over one week in July for a 10 kV system with a system load of 8.7 MW. It relates to a purely residential area on the outskirts of a city.

In Figure 2.12, the course of the fifth harmonic voltage from Monday to Friday morning is identical. The daily time-course which was shown in Figure 2.11 and discussed in the introduction is reflected here. The level of the fifth harmonic voltage increases to a maximum and the curve trace is flatter at the weekend. The increase in the maximum levels on both Saturday and Sunday afternoon can clearly be seen, and are attributable to altered user behaviours (television sets). The increase in the level of harmonics at the weekend can also be attributed to the reduced loading on the supply network, which affects the attenuation of harmonic voltages.

2.1.5 Telecontrol signals

With telecontrol systems, control signals are transmitted via the mains to telecontrol receivers; e.g. for tariff meter switching, for lighting control or for alerting personnel. Older systems predominantly operate in the frequency range of 110 Hz to 3 kHz, whilst modern systems operate in the range of 110 to 500 Hz. The operating frequency in the range below 500 Hz lies mainly between the typical harmonics, and in the range above 500 Hz, at harmonic frequencies

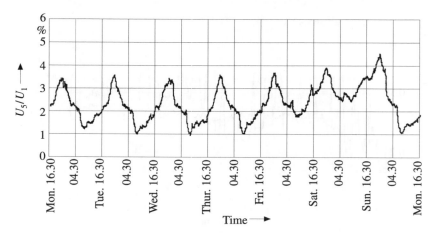

Figure 2.12   Time course of 5th harmonic (linear scale) in a 10 kV system
Measured from 11/7 till 18/7/94; system load 8.7 MW; residential area

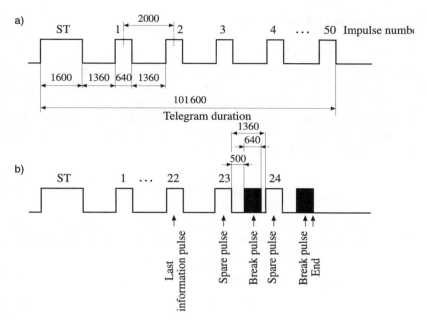

Figure 2.13   Pulse characteristic of mains control telegram; times in ms
a) fixed length of telegram
b) variable length of telegram

which are not generated by three-phase bridge circuits in steady-state condition. Telecontrol signals are transmitted as short-duration impulse telegrams containing the relevant telecontrol frequencies. The total duration of the telegram is about one minute. Figure 2.13 shows two examples of telecontrol signals (the r.m.s. value of the telecontrol voltage is shown). Telecontrol signals should be considered as harmonics or interharmonics in respect of perturbations depending on the telecontrol frequency. The relationship of the functional voltage and control voltage in respect of the transmitter frequency f_s is defined in EN 61037 (VDE 0420 part 1: 1994–01). The control voltage U_{max}, is the impressed voltage of the telecontrol transmitter in the supply, and the functional voltage U_f the voltage of the telecontrol receiver with which it communicates. The following relationships apply:

$$\frac{U_{max}}{U_f} \geq 8 \quad \text{to} \quad f_s > 250 \text{ Hz} \tag{2.20a}$$

$$U_{max} \geq U_f \left(8 + \frac{(f_s - 250) \cdot 7}{500} \right) \quad \text{to} \quad f_s = 250 \text{ Hz to } 750 \text{ Hz} \tag{2.20b}$$

$$\frac{U_{max}}{U_f} \geq 15 \quad \text{to} \quad f_s > 750 \text{ Hz} \tag{2.20c}$$

2.2 Description and calculations

2.2.1 Characteristics and parameters

Because of the physical relationships, active power can only be generated between currents and voltages of equal frequency. Harmonic currents can only convert alternating powers with voltages of other frequencies and thus also with the fundamental component of voltage. Assuming that the voltage is purely sinusoidal, the apparent power of a current containing harmonics and a sinusoidal voltage is calculated according to

$$S^2 = U^2\left(I_{w1}^2 + I_{b1}^2 + \sum_{h=1}^{\infty} I_h^2\right)$$
(2.21)

with the active power P_1 and the reactive power Q_1 of the fundamental component of current and the distortion content D of the current harmonics according to

$$P_1 = U I_1 \cos \varphi$$
(2.22a)

$$Q_1 = U I_1 \sin \varphi$$
(2.22b)

$$D = U\sqrt{\sum_{h=1}^{\infty} I_h^2}$$
(2.22c)

The quantities can be represented in a right-angled system of co-ordinates as shown in Figure 2.14.

If voltages and currents are not sinusoidal, it must be noted that active power is also converted by the harmonics of equal frequency in current and voltage. (See also section 1.4.4.)

The following definitions (valid for currents and voltages) of relative values, shown here in Equation (2.23a) to Equation (2.23c) using current as an example are determined as follows, according to DIN 40110:

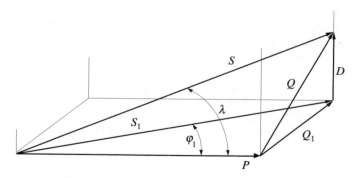

Figure 2.14   Vector diagram of different parameters of electrical power in a.c. systems according to DIN 40110

- *r.m.s. value I* as a root of the quadratic sum of the harmonic currents.

$$I = \sqrt{\sum_{h=1}^{\infty} I_h^2}$$ (2.23a)

- *Fundamental component content g* as a quotient of the r.m.s. value of the fundamental component to the total r.m.s. value.

$$g = I_1/I$$ (2.23b)

- *Harmonic content k* or harmonic distortion factor as a quotient of the r.m.s. value of the harmonics of the total r.m.s. value.

$$k = \frac{\sqrt{I^2 - I_1^2}}{I} = \sqrt{1 - g^2}$$ (2.23c)

The *THD* (*total harmonic distortion*) is not defined in DIN 40110 and is calculated as a quotient of the r.m.s. value of the harmonics relative to the fundamental component r.m.s. value.

$$THD = \frac{\sqrt{I^2 - I_1^2}}{I_1} = \frac{k}{g} = \sqrt{\sum_{n=2}^{40} (I_h/I_1)^2}$$ (2.23d)

To assess the harmonics of certain orders, *THD* weighting factors can be introduced into the calculation of the harmonic distortion (see draft IEC 1000–3–4). The characteristics determined in this way are known as the *partial weighted harmonic distortion* (PWHD).

$$PWHD = \sqrt{\sum_{n=14}^{40} h(I_h/I_1)^2}$$ (2.23e)

Despite the absence of a definition in DIN 40110, the *THD* and not the harmonic content *k* (previously harmonic distortion factor) are currently used.

The *power factor λ* as a quotient of the active power and the apparent power applies generally for non-sinusoidal currents and voltages according to

$$\lambda = \frac{P}{\sqrt{(P^2 + Q_1^2 + D^2)}}$$ (2.23f)

The *displacement factor* cos φ_1 as a quotient of the active power relative to the fundamental apparent power is defined as the fundamental power factor in the case of sinusoidal voltage and non-sinusoidal currents.

$$\cos \varphi_1 = \frac{P}{\sqrt{(P^2 + Q_1^2)}}$$ (2.23g)

The power factor and displacement factor quantities are shown in addition to the power quantities in Figure 2.14.

On the basis of the *THD* of the voltage, which is also known as the *distortion factor d*,

$$d = \sqrt{\sum_{h=2}^{n}\left(\frac{U_h}{U_1}\right)^2} \tag{2.24a}$$

the distortion factors d_{ind} and d_{cap} are calculated according to Equation (2.24b) and Equation (2.24c) to estimate the effects of the harmonics on inductances and capacitances.

$$d_L = \sqrt{\sum_{h=2}^{n}\left(\frac{U_h}{h^a U_1}\right)^2} \tag{2.24b}$$

$$d_C = \sqrt{\sum_{h=2}^{n}\left(h\frac{U_h}{U_1}\right)^2} \tag{2.24c}$$

The various iron-core qualities are taken into account by the exponent α, which is usually between 1.5 and 3.

To describe the superimposition of harmonic currents from various causes, the *co-phasal factor* k_{ph} according to Equation (2.25) is defined as the quotient from the geometric to arithmetical sum of the currents under consideration. According to the definition, k_{ph} is always ≤ 1.

$$k_{ph} = \frac{\left|\sum_{base}^{top} I_h\right|}{\sum_{base}^{top} |I_h|} \tag{2.25}$$

2.3 Harmonics and interharmonics in networks

2.3.1 Calculation of networks and equipment

Calculations of harmonics and interharmonics in electrical networks may be performed to analyse disturbances, to plan and design compensation systems or to calculate the propagation of telecontrol signals. It is assumed that the system is in a steady state. In this case calculation methods can be used in both the time and frequency range, see also [14, 15].

For analyses in the time range, the system state is determined by the node voltages and branch currents, the relationship of which is described using a system of differential equations. This can be solved by the normal numerical methods. The result is the calculated time course of current and voltage in discrete time intervals. The method enables all the processes in a network to be calculated, including those of the controller. Non-linearities in the equipment and consumers can be allowed for. To determine the harmonics in the steady state, the time courses up to decay of the transient reactions must be calculated. The harmonic content can then be determined with the aid of Fourier analysis. To justify the high modelling cost, the long calculation times and the demand on memory capacity, methods are used in the time range, preferably to calculate transient occurrences in physically small networks with a small number of converters.

If it is necessary to calculate the steady state harmonics in extended networks, the calculation method is used in the frequency range. For this purpose, the differential equation system (time range) is converted to a complex algebraic equation system (frequency range). The harmonics in the frequency range can be represented by complex phasors, which can be described by the amount and phase angle or by real and imaginary parts. There is also an analogy here to the considerations of the Fourier analysis in section 1.4.2.

In the following, a closer consideration is given to the process of linear harmonic analysis, where the required data can be taken from the name plate data of the equipment in the same way as for load flow and short-circuit calculations. Perturbations of non-linear consumers with each other, and non-linear effects, such as iron-core saturation of transformers, cannot be simulated by a process of linear harmonic analysis. The harmonic currents of non-linear consumers are considered as constant impressed currents. This assumption is justified because the harmonic currents in the area of the usual distortion factors (harmonic content) of the voltage is almost independent of the voltage waveform.

2.3.2 Modelling of equipment

The transmission behaviour of equipment and loads is linearly modelled and described by the node admittance matrix, which must be calculated separately for each frequency to be considered. Three-phase calculations are carried out in three-phase components, as well as single-phase in symmetrical components. To determine harmonics in electrical power supply systems, it is generally sufficient to carry out single-phase calculations in symmetrical components and to model positive sequence, negative sequence or zero sequence systems, depending on the order of the harmonics to be determined, or the direction of rotation of inter-harmonics. The voltage harmonic content is then calculated on the basis of the impressed harmonic currents.

The equipment should be modelled from the available characteristic data, which is also necessary for other network calculations. Calculations in the range of interest up to the 40th harmonic should be possible, with an increased modelling accuracy being sought in the frequency range up to about 1 kHz. Cables, overhead lines and transformers are simulated by π equivalent circuits (conventional or based on the line equations). In contrast to the T equivalent circuit, no new nodes occur.

The equivalent circuits for cables and overhead lines take account of the quantity per unit length of capacitance, inductance and resistance, which can be determined from the conductor cross-section, arrangement and material and also the type of insulation. The conventional π equivalent circuit can be used for overhead lines up to 250 km long, for cables up to 150 km, divided by the harmonic order h. The accuracy of the modelling in this case reduces with the increase in frequency and line length. If an increased accuracy is required, the line must be divided into sections and the individual π equivalent circuits must be connected in series. The π equivalent circuit based on the line equations, which describes the transmission property of a line without additional modelling expense, is better suited to investigating harmonics. The normal lines up to approximately 2 km in low and medium voltage networks are described in the frequency range up to 1 kHz by the conventional π equivalent circuit.

Transformers are also modelled by a π equivalent circuit with ideal transformer ratio. The parameters of the equivalent circuit are determined from the vector group, transformer ratio and the quantities determined from the short-circuit and open-circuit measurements. Because the natural resonant frequencies of transformers are above 5 kHz and the winding capacitances are relatively small compared to line capacitances, the winding capacitances are not simulated. The vector group and phase rotation of transformers should be considered with regard to transmission of harmonics over different network levels.

Generators, motors and the network supply represent consumers for harmonics investigations whose 50 Hz source voltages are considered as short-circuited. The equivalent circuits are based on the (subtransient) short-circuit data.

To correctly simulate possible resonances of superimposed network levels, it is necessary to represent these by a parallel resonant circuit, which contains the short-circuit impedance of the network supply, the sum of the distributed line and compensation capacitances, as well as the resistance resulting from the active load. Pre-emphasis of the voltage in superimposed network levels is simulated by equivalent current sources or equivalent voltage sources of a corresponding frequency.

Linear active loads represent the attenuating components of the network, which can be simulated as a close approximation by a purely ohmic resistance corresponding to the active power content at the load. Inductive loads can be represented by a parallel inductance according to the reactive power content of the load. The capacitive content superimposed by the inductive fundamental reactive power cannot usually be determined from the load data. It can be estimated using load factors based on operating experience.

2.3.3 Resonances in electrical networks

If one assumes, when considering the effect of harmonics and interharmonics in electrical power systems, that during the operation of equipment generating harmonics or interharmonics the voltage harmonics produced at the connection point are of interest, this problem can be usually reduced to a simple structure as shown in Figure 2.15. Because the processes for harmonics and interharmonics in this investigation are identical, the harmonics are considered in the following by way of example. The same applies in each case for the interharmonics.

Generally, such a power supply structure consists of a supply via a transformer from a network with a higher voltage level. At the connection point, or point of common coupling, loads other than those generating harmonics, such

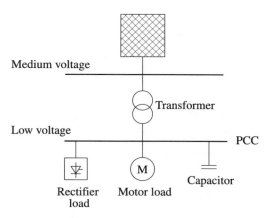

Figure 2.15   *Power system diagram indicating a simplified structure of electrical power supply scheme to industrial consumers*

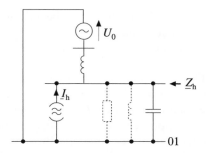

Figure 2.16   *Electrical diagram for the power system. See also Figure 2.15*

as ohmic and motor consumers, are connected. A capacitor bank is often used for reactive power compensation. The aforementioned ohmic and motor consumers are sometimes connected to the point of common coupling by cables. The cable capacitances and the capacitors must be taken into account in the investigation.

For a further investigation of the processes with regard to harmonics, the equivalent circuit of the network in the positive sequence system, shown in Figure 2.16, is used. It is known that the inductive supply and the capacitive reactive current compensation, or the cable capacitances, from the point of view of the harmonics generator, form a parallel resonant circuit at the point of common coupling, which is attenuated by the ohmic content of the supply and of the loads. Equation (2.26) is often used to supplement the calculation of the resonant frequency according to Equation (1.41),

$$f_{res} = f_1 \sqrt{S_r/(u_k Q_C)}, \tag{2.26}$$

with the rated apparent power S_r and short-circuit voltage u_k of the supplying transformer and the rated power Q_C of the capacitor bank.

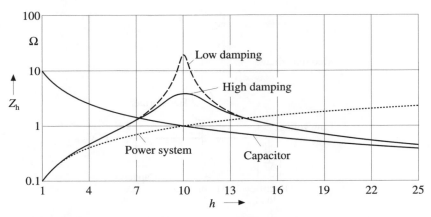

Figure 2.17   Impedance versus harmonic order at the point of common coupling (PCC) of the power system. See also Figure 2.15

The effects of the current harmonics can be calculated after this initial investigation. The equations determined in section 1.4.5 are used for the description. The course of the impedance at the point of common coupling is shown in Figure 2.17.

The impedance of the parallel resonant circuit increases, starting from the impedance of the inductive supply in the case of low frequencies, to the maximum value at the resonant frequency f_{res}, but again reduces with further increases in frequency and approaches the impedance of the capacitive part. The impedance at the resonant point is equal to the impedance of the supply multiplied by the quality Q or divided by the attenuation $\underline{d}$.

$$|Z_{res}| = \frac{\omega L}{\underline{d}} \qquad (2.27)$$

If one assumes that the harmonic currents are impressed currents, these coincide with an impedance value, which is increased compared with the impedance of the supply or compared with the impedance of the capacitances, in the frequency range

$$\frac{f_{res}}{\sqrt{2}} < f < f_{res}\sqrt{2} \qquad (2.28)$$

and thus also result in increased voltages. The motor loads, represented by their inductivity, lead to a shift in the resonant frequency to lower frequencies. However, this effect is relatively slight, taking account of the impedance values of the supply and the motor loads.

Substantially greater effects on the impedance at the system connection point are caused by a change in the capacitor rating, e.g. by a stepped capacitor bank. Figure 2.18 shows the change in the impedance course where the capacitor bank can be switched in steps from $Q_C = 100$ kvar to 550 kvar. In the network under

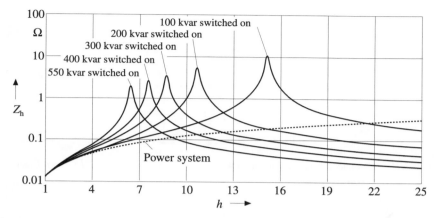

Figure 2.18 Impedance versus harmonic order at the point of common coupling (PCC) of an industrial system
$S''_k = 23.8\ MVA$; switched capacitors $Q_C = 100\ kvar \ldots 550\ kvar$ (5 steps)

consideration (short-circuit power $S''_k = 23.8$ MVA at the point of common coupling), the resonant frequency changes in the $f_{res} = 304$ Hz to 780 Hz range. Considering the impedance increase according to Equation (2.28), resonance-related voltage increases must be expected in the $f = 214$ Hz to 1103 Hz range (harmonic orders $h = 5$ to 22).

The voltage increase at harmonic frequencies leads to a high current loading of the capacitors,

$$I_{Ch} = U_h\, h\, \omega_1 C, \qquad (2.29)$$

because the impedance of the capacitor drops with rising frequency. These currents can sometimes be greater than the impressed harmonic currents and cause damage to the capacitor.

Up to now the impedance of the power system has been considered from the point of view of the harmonics generator at the point of common coupling with impressed harmonic currents, in the following it is considered from the point of view of the supplying network. Figure 2.19 shows the equivalent circuit in the positive sequence system of the power system according to Figure 2.15.

Inductive impedance of the supplying transformer and capacitive impedance of the capacitor are now in series and form a series resonant circuit which is attenuated by the ohmic content. The course of the impedance is shown in Figure 2.20.

The impedance of the series resonant circuit drops, starting from the impedance of the capacitor at low frequencies, to the minimum value at the resonant frequency, but rises again with a further increase in frequency and approaches the impedance of the reactance of the supply. The impedance at the resonant point is equal to the impedance of the supply, divided by the quality Q or multiplied by the attenuation A.

$$|Z_{res}| = \omega L d \qquad (2.30)$$

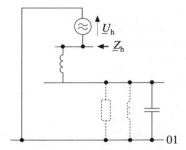

Figure 2.19   Electrical diagram for the power system (see also Figure 2.15) seen from HV connection

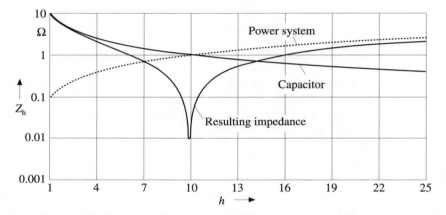

Figure 2.20   Impedance versus harmonic order at HV connection of the power system; see also Figure 2.19

Even where there are small voltage harmonics in the supplying networks, large harmonic currents flow through the transformer into the capacitor system and can also damage the capacitors.

The influence on the resonant frequency of a compensating system switched in steps is similar to the examination of the parallel resonant effect.

2.4 Effects of harmonics and interharmonics

2.4.1 General

Because of the impedance relationships in electrical networks, the current harmonics from secondary networks can be regarded as impressed source currents and the voltage harmonics from primary networks can be regarded as impressed source voltages (see also section 1.5.1). In this case, the harmonics superimpose on each other vectorially. Because the third harmonic and its multiples form zero phase-sequence systems, these do not generally pass from the low voltage network to the superimposed medium voltage network, because the zero sequence system cannot be transmitted due to the vector group and earthing of the supplying transformers (Dy or Dz). Because of the actual finite zero impedance of the delta windings, or of the non-earthed windings in the neutral point, up to a maximum of 20% of the harmonics of the zero sequence system is transferred into the superimposed voltage level.

2.4.2 High-energy equipment

On three-phase a.c. motors and generators, current harmonics cause additional temperature rise and develop disturbing moments similar to the fundamental components of current of the negative sequence system I_{21} during starting. For this reason, the total r.m.s. value of the current harmonics I_h and fundamental components of current I_{21} of the negative sequence system produced at the motor short-circuit inductance by the voltage harmonics, as per

$$I = \sqrt{I^2_{21} + \sum_{h=1}^{\infty} I^2_h} \tag{2.31}$$

according to EN 60034–1 (VDE 0530 part 1:1995–11), Table 7 may not exceed

$$I = \sqrt{U^2_{21} + \sum_{h=1}^{\infty} \left(\frac{U_h}{h}\right)^2 \cdot \frac{I_{an}}{U_1}} \tag{2.32a}$$

$$I \le (0.05 \ldots 0.1)I_{rM} \tag{2.32b}$$

In this case small values apply for directly-cooled machines and for machines with a greater rating (up to 1.6 MVA), large values are permissible for indirectly-cooled motors. The torque M of asynchronous motors is proportional to the square of the r.m.s. values of the stator voltage U,

$$M \sim \frac{U^2}{n_1 X_\sigma}, \tag{2.33}$$

and for synchronous machines is proportional to the stator voltage U as follows:

$$M \sim U\frac{U_p \sin \vartheta}{n_1 X_d}, \tag{2.34}$$

Harmonics in the voltage result in synchronous or counter-rotating torques depending on the order (see section 1.4.3) of the harmonic.

Multiples of the third order form zero sequence systems but no torques, because these are pure alternating fields. The higher-frequency rotating fields lead to uneven running of machines due to the higher-frequency torques, which

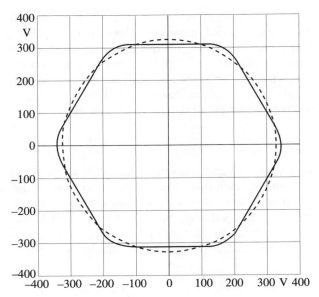

Figure 2.21   *Voltage space vector in an LV system*
———— *basic frequency*
— — — — *basic frequency and 5th harmonic $U_5/U_1 = 0.05$*

has the effect of disturbing noise and vibrating moments. Oscillations can, under certain circumstances, also be induced between the individual masses on the generator or motor shaft.

Figure 2.21 shows the course of the space vector of the voltage for a low - voltage network, where the amount of the fifth voltage harmonic U_5 is equal to 5% of the fundamental component of voltage U_1. The change frequency of the voltage rotation vector is $6 \times f_1$, i.e. equal to the difference between the fundamental component frequency (positive sequence system) and the fifth harmonic frequency (negative sequence system).

For capacitors, the total r.m.s. value of the current caused by the voltage harmonics, according to

$$I = \omega_1 C \sqrt{\sum_{h=1}^{\infty} (hU_h)^2} \qquad (2.35)$$

as per EN 60831–1 (VDE 0560 Part 46: 1997–12), must not exceed 1.3 times the rated current, or 1.5 times the rated current if capacitance tolerance of $1.15 \times C_n$ is assumed.

Furthermore the rise in the dielectric losses

$$P_d = U_1^2 h \, \omega_1 C \tan \delta, \qquad (2.36)$$

which increase in proportion to the square of the voltage, should also be noted.

Otherwise the following applies for the voltage.

$$\left(\frac{U}{U_n}\right)^2 + \sum h\left(\frac{U_h}{U_n}\right)^2 \leq 1.44 \qquad (2.37)$$

The maximal permissible voltage, according to EN 60831–1 (VDE 0560), depends on the duration of the voltage stress, as given in Table 2.3.

Table 2.3 Permissible voltage stress of capacitors relative to the stress duration

Voltage U_{max}	Time T_{max}
$U \leq 1.0 \times U_n$	Permanent
$U \leq 1.1 \times U_n$	8 hours/day
$U \leq 1.15 \times U_n$	30 minutes/day
$U \leq 1.2 \times U_n$	5 minutes/day
$U \leq 1.3 \times U_n$	1 minute/day

Lines (overhead wires, cables and rails) experience a higher stress due to harmonics, depending on the frequency and therefore a local temperature rise which is increasingly pronounced from about 1000 Hz due to the skin effect. In networks with a fourth conductor (low voltage system types TN and TT according to IEC 364–3 (VDE 0100 Part 300)) as a return conductor this can lead, where harmonic components are present in the voltage, to a current with a frequency of 150 Hz in the neutral conductor. Where there is a high degree of non-linear consumers, higher stress of the neutral conductor compared with the phase conductors can occur. Figure 2.22 shows, as an example, the frequency spectrum of the current in a low voltage system that almost exclusively supplies PCs, displays and compact fluorescent lamps in an office building.

Where there is no neutral conductor, a displacement voltage with respect to earth with a corresponding frequency forms in the neutral point of the network. Sometimes currents containing harmonics have a larger di/dt at the zero crossing than a corresponding sinusoidal current with an equal r.m.s. or peak value. This can reduce the quenching capability of circuit breakers. Vacuum circuit breakers are less susceptible to this than magnetically-blown switches. Fuses are generally less susceptible to harmonics, where only a premature tripping occurs relative to the rated value, which in view of the additional temperature rise of the equipment to be protected can be very desirable.

With transformers, operation at non-sinusoidal voltage and/or with non-sinusoidal current leads to increased ohmic losses and also to a rise in the eddy current losses and hysteresis losses. Monitoring the current loading of the transformer compensating winding can also be problematic (delta circuit), if only the current in the neutral winding is measured, which would mean that the content of the third harmonic would not be detected.

Inductive voltage transformers can become saturated due to harmonics, thus substantially increasing the transformation error. Current transformers are

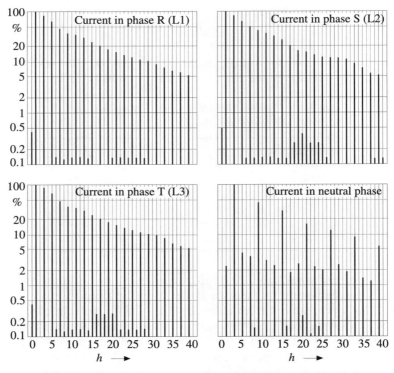

Figure 2.22   RMS-values of harmonics (related to highest harmonic) of load current in a TN-system (0.4 kV), line currents in conductors L1, L2, L3, current in neutral conductor N

generally less sensitive in this case, with only the phase-angle error being detrimentally affected. This needs to be considered for measuring harmonics.

2.4.3 Network operation

Medium voltage systems (6 kV to 30 kV, sometimes up to 110 kV) are often operated with earth-fault compensation (Petersen coil). This means that the inductance of the earth-fault compensation coil is set almost in resonance with the conductor-earth capacitances of the system. The earth-fault residual current I_{Rest} flowing through the fault point is very small and, due to overcompensation, is inclined to become ohmic-inductive. This makes the quenching of the earth fault considerably easier in the case of faults in air. The harmonic content of the current occurring at the earth fault point due to the presence of harmonics in the voltage can not be compensated for and becomes superimposed on the supply-frequency earth-fault residual current, as follows:

$$I_{Rest} \approx j\left(\sqrt{3}\,U_n\omega C_E\left(1 - \frac{1}{\omega^2 L_{OD}C_E}\right)\right) \tag{2.38}$$

VDE 0228 Part 2:1987–12 stipulates limits for the quenching capacity of ohmic earth-fault residual currents I_{Rest} and capacitive earth-fault currents I_{CE}. For systems where $U_n = 20$ kV these amount to:

$$I_{Rest} \leq 60 \text{ A and } I_{CE} \leq 36 \text{ A}$$

The r.m.s. value of the displacement voltage at the arc-suppression coil is used for automatic tuning of the Petersen coil. This is changed by the voltage harmonics in the system and correct tuning of the Petersen coil is therefore hampered.

2.4.4 Electronic equipment

Electronic equipment can be affected not only by voltage harmonics but also by other system perturbations to the extent that proper functioning is impaired or the equipment is damaged.

Causes of the effects due to harmonics are the shifting of the zero crossings and the occurrence of multiple zero crossings. Because of this malfunctions can occur on equipment which has to detect zero crossings of voltage, e.g. in converter control systems, synchronising devices and equipment for parallel switching. Sometimes the cause of the disturbance and the disturbed consumer can be one and the same.

The proper functioning of telecontrol receivers, which nowadays are designed as electronic equipment, can be impaired by harmonics or interharmonics if the harmonic level exceeds the limits stipulated in EN 61037 (VDE 0420). The limits are all above the compatibility levels stated in the different parts of EN 61000 (VDE 0839).

The propagation of telecontrol signals, and thus the correct response of telecontrol receivers, depends on the particular network impedances. In particular, the interaction of impedances of inductive supplies and capacitive network parts such as capacitors for reactive power compensation (which from the point of view of the input telecontrol signal form series resonance circuits) must be considered to ensure an adequate signal level at the customer system. Because these are network-specific processes which are a direct consequence of 'interharmonic' or 'harmonic' system perturbations, this phenomenon is not dealt with further in this book. For further information, see [16]

2.4.5 Protection, measuring and automation equipment

The effect of system perturbations on protective equipment such as distance protective devices, overcurrent protective devices and differential protective devices, depends heavily on the construction and operation of the equipment. Data and information from the manufacturer are necessary for planning systems and tracing disturbances. The effect on triggering devices for low voltage and medium voltage circuit breakers is given in the following as an example.

Analogue triggering devices for overload protection are particularly vulnerable to harmonics because the current-proportional voltage course $u(t)$ after the peak value filter depends only on the peak value of the current I. The peak value is in a defined relationship to the r.m.s. value only with sinusoidal current. Figure 2.24a is a block diagram of an analogue tripping device and also shows current and signals. For the sinusoidal current course shown in Figure 2.23a, the tripping device can be set precisely to a defined r.m.s. value of current which causes it to trip if overshot.

In the case of the non-sinusoidal current shown in Figure 2.23b, the content of the third harmonic was chosen so that the peak value of the total current is less than the peak value of the fundamental component according to Figure 2.23a. According to Equation (2.23a), the total r.m.s. value of the current applied to the equipment is greater. The peak value filter determines a voltage proportional to the peak value of the current, which as a measure of the r.m.s. value of the associated current represents a value which is too small. In this case the tripping device would not trip, which would mean that the equipment to be protected would be subjected to excessive stress under certain circumstances.

This tendency of analogue tripping devices to malfunction can be rectified by using digital tripping devices. In this case, the r.m.s. value of the voltage proportional to the current is formed by sampling the rectified measurement signal. To do this, it is sufficient to sample the measuring signal at approximately 1 kHz. The tripping of the A/D converter must not exceed 12 bit. Figure 2.24b is a block diagram of a digital tripping device.

The influence of harmonics on the accuracy of induction meters is considerable. Harmonics can also cause mechanical oscillations because the natural frequency is in the $f_{res} = 400$ Hz to 1000 Hz range. Electronic meters should be used in systems with a high harmonic content. Their accuracy depends mainly on the sampling frequency used and the resolution accuracy. Measuring instruments for other purposes should be checked for suitability of use with non-sinusoidal quantities.

2.4.6 Loads and consumers

Harmonics shorten the service life of lamps by increasing the filament temperature. In the case of fluorescent lamps and other gas discharge lamps, harmonics can lead to a disturbing level of noise. It should also be noted that fluorescent lamps are often fitted with capacitors for reactive power compensation. In this case, the effect of the overloading of the capacitors (see section 2.4.2) should be noted. Furthermore the capacitors, together with the inductive load, form a resonant circuit. The resonant frequency for individual compensation is a maximum of 80 Hz, and thus resonant excitation is not expected [6]. For group compensation the resonant frequencies can sometimes be higher. This must be considered on a case-for-case basis in the planning phase.

Disturbances of power equipment and information technology equipment can cause secondary damage in industrial systems. In this case, the uncontrolled shutdown of equipment and production processes must be considered, as this

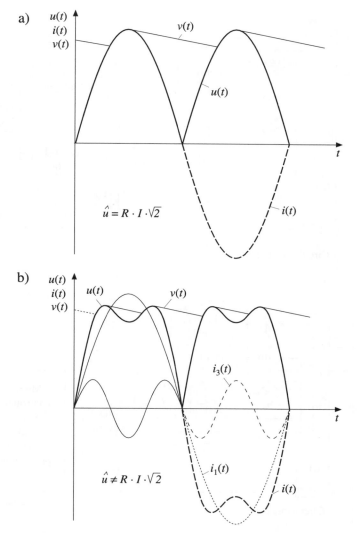

Figure 2.23   *Current and voltage signals in an analogue cut-out*
     a) sinusoidal load current; signals
     b) consumer current with 3ʳᵈ harmonic; signals

secondary damage can often be many times greater than the cost of counter-measures to reduce harmonics.

Where the distance between overhead lines and telephone lines is small, speech transmission can be distorted. The human ear is most sensitive in the 1 kHz to 1.5 kHz range. Particular attention must therefore be paid to harmonics in the 20th to 30th order. These cause inductive, capacitive and galvanic couplings (local increase in the reference potential).

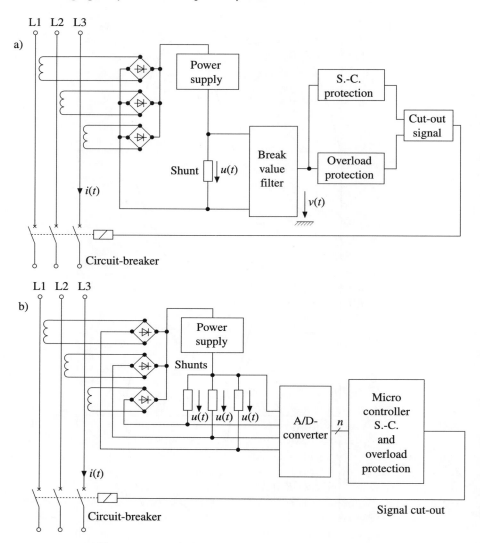

Figure 2.24   a) Block diagram of an analogue cut-out
b) Block diagram of a digital cut-out

A psophometric weighting of the various current and voltage harmonics by the telephone interference factor *TIF* is used to assess the harmonics:

$$TIF = \frac{\sqrt{\sum_{0=0}^{n}(k_{fe}C_h)^2}}{C_1} \tag{2.39}$$

using the weighting factor

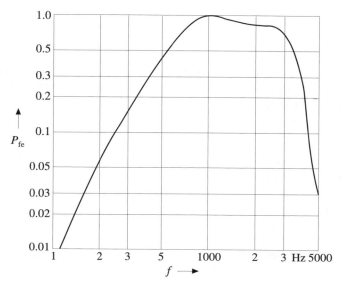

Figure 2.25   Weighting factor k_{fe} for determination of TIF (telephone interference factor)

$$k_{fe} = P_{fe}\, 5f_h;\tag{2.40}$$

P_{fe} (see Figure 2.25).

2.4.7 Assessment of harmonics

Not only the maximum value but also statistical characteristics such as the 95% and 99% value of the frequency are decisive in assessing the harmonics problem. Cumulative frequencies, mean values and the standard deviation must also be considered. These values are calculated by modern measuring systems for system perturbations and form the basis of further assessments and, if necessary, the stipulation of remedial measures. Measurements are also sometimes required to be evaluated over a period of several weeks. It may also be necessary to differentiate between work days and weekends (see section 7.2).

As explained in section 1.5.1, current harmonics from secondary networks act as impressed currents, while voltage harmonics from primary networks act as impressed voltages. Thus it is understandable that each network level (low - voltage, medium voltage and high voltage) can only be assigned to that part of the particular compatibility level which corresponds to its amount of the total network impedance, according to section 1.5.1, over all voltage levels. This is expressed by the system level factors k_N according to [7]. For low-, medium and high voltage networks, these are in the following ranges:

$$k_{NNS} : k_{NMS} : k_{NHS} \approx (0.2 \ldots 0.3):(0.4 \ldots 0.7):(0.1 \ldots 0.3)\tag{2.41}$$

The harmonic current permissible in a network in each case is calculated as follows:

$$I_{hmax} = k_N \frac{U_{hVT}}{Z_h} \qquad (2.42)$$

The frequency-dependent network impedance Z_h from the short-circuit react-ance X_h of the network supply or from the initial symmetrical short-circuit a.c. power S''_{kV} at the point of common coupling V and the impedance angle ψ can be calculated as follows:

$$Z_h \approx h \frac{U_n^2 \sin \psi}{S''_{kV}} \qquad (2.43)$$

In this simplified approach, Z_h does not take account of all network resonances and can therefore lead to incorrect decisions in some cases. For instance, the frequency-dependent impedances of a 10 kV municipal network for various load situations are given in Figure 2.26 [8, 9].

When assessing harmonics, it should be noted that the harmonic currents generated by various pieces of equipment are added corresponding to their phase angle. This is described by the co-phasal factor k_{ph} according to Equation (2.25) (quotient of the geometrical to the arithmetical sum), as follows:

$$k_{ph} = \frac{\left| \sum\limits_{base}^{top} \underline{I}_h \right|}{\sum\limits_{base}^{top} |\underline{I}_h|} \qquad (2.25)$$

If one also considers that several harmonic generators are normally con-nected in the network, each consumer i can only be assigned the content at the complete compatibility level U_{hVT} which corresponds to its content S_i of the complete load of the network S_N or at the output of the feeding transformer S_{rT}. This is described by the system connection factor k_A:

$$k_A = S_i/S_N \qquad (2.44a)$$

or

$$k_A = S_i/S_{rT} \qquad (2.44b)$$

The maximum permissible harmonic current I_{hmaxi} of a consumer i is therefore calculated as follows:

$$I_{hmaxi} = \frac{k_A k_N U_{hVT} S''_{kV}}{k_{ph} h U_n^2 \sin \psi} \qquad (2.45a)$$

or

$$I_{hmaxi} = \frac{k_A k_N U_{hVT}}{k_{ph} Z_h} \qquad (2.45b)$$

The permissibility of the connection or of the operation of harmonics gener-ators can also be suitably estimated by considering the harmonic distortion factors B_h. In the case of a separate equipment i, the generated relative harmonic voltage u_{hi} is used as a basis for the calculation, as follows:

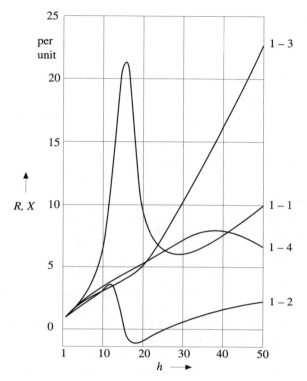

Figure 2.26 Frequency dependency of the impedance of a 10-kV system
1–1 resistance at peak load (daytime)
1–2 reactance at peak load (daytime)
1–3 resistance at low load (nighttime)
1–4 reactance at low load (nighttime)

$$u_{hi} = Z_h \frac{I_{hi}}{U_n/\sqrt{3}} \tag{2.46}$$

If the network is supplying several converters, the resulting harmonic voltage is calculated as follows:

$$u_h = z_h \sum k_{\text{ph}i} i_{hi} S_{ri}, \tag{2.47}$$

whereby z_h in %/MVA and i_{hi} are given as relative harmonic currents. If the harmonic voltages caused are considered relative to the compatibility level $u_{h\text{VT}}$, harmonic distortion factors B_h are obtained

$$B_h = \frac{z_h \sum\limits_{0=0}^{n} k_{\text{ph}i} i_{hi} S_{ri}}{u_{h\text{VT}}} \tag{2.48}$$

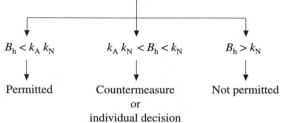

Calculation of B_h according to Equation (2.47) and Equation (2.48)

$B_h < k_A k_N$	$k_A k_N < B_h < k_N$	$B_h > k_N$
Permitted	Countermeasure or individual decision	Not permitted

Figure 2.27   Decision tree for assessment of harmonics for a single consumer

The unrestricted connection and operation of the system or equipment is then permissible only if the current harmonics are less than the amounts attributable to this equipment, taking account of the system connection factor k_A and system level factor k_N.

The associated harmonic distortion factors B_h must then be less than the product of the system level factor k_N and system connection factor k_A. If the harmonic distortion factor B_h is greater than the system level factor k_N, this means that the overall effective level of the voltage harmonics U_h for this harmonic order h is greater than the level assigned to this network level. Neither connection nor operation are thus permissible (see also Figure 2.27). For all other cases, the permissibility of the connection and operation, as well as suitable countermeasures, must be decided separately for each case.

2.5 Standardisation

2.5.1 General

It was explained in section 1.1 that, when considering system perturbations in general and harmonics in particular, a balance must be struck between the economic needs and technical boundaries of both the consumer and network operator. The standardisation must take account of these aspects and therefore also offers various approaches for achieving this aim. If the investigation is limited to conducted system perturbations to which the harmonics and inter-harmonics are attributable, the galvanic couplings between interference emitting equipment and disturbed equipment must be examined. There are three main courses of action in this case:

- limiting the emitted interference,
- reducing the coupling between the disturbing and disturbed equipment,
- increasing the interference immunity.

For this, the individual disturbance phenomena must be considered separately.

2.5.2 *Emitted interference*

Measurements of harmonic levels in the last 20 years have revealed steadily-increasing levels [10]. For example, the level of the fifth voltage harmonic in the networks of Germany have doubled to 6%. This means that the compatibility levels are now exceeded in some networks. However, it cannot be raised because the interference immunity of the equipment operated in the network is aligned with it. Limitation of the emitted interference is therefore an urgent task.

EN 61000 Part 3–2 (VDE 0838 Part 2): 1996–03 stipulates limits for the emission of current harmonics from equipment with input currents per conductor of $I_1 < 16$ A, with the equipment being divided into the following four classes.

A: Symmetrical three-phase equipment and all other equipment, apart from those assigned to classes B, C or D.
B: Portable electrical tools (1.5 times the class A limit).
C: Lighting equipment, including dimmers.
D: Equipment with an input power of $P = 50$ W up to 600 W and the current course, according to Figure 2.28, determined under specified test conditions.

Standardisation for equipment with fundamental components of current of more than 16 A are presently under discussion [10]. In this case it should be particularly noted that the interference emission limits coincide with the limits of existing equipment, the interference emission limits of which are stipulated in EN 61000–3–2 (VDE 0838 Part 2). The limits for class B (portable electric tools) should be chosen for this purpose.

A multistage acceptance procedure is proposed here. Stage 1 in this case covers equipment where the short-circuit power S''_k at the point of common coupling (PCC) must be at least 33 times greater than the equipment power S_G. This would result in a maximum voltage drop of 3% at power system frequency.

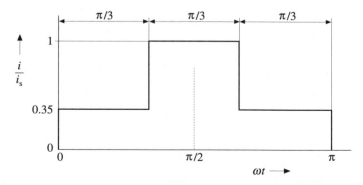

Figure 2.28   Envelope of the input current to define the 'special wave shape' and to classify equipment as per class D (EN 61000–3–2)

This means, for instance, that the value of $I_5/I_1 = 10.7\%$ proposed for the fifth harmonic (identical to the class B limits), taking account of typical summation factors, would not lead to the compatibility levels being exceeded if this value for the fifth harmonic is due to approximately 55% of consumers. Table 2.4 shows the proposed interference emittance limits for harmonic currents, according to IEC 1000 3–4.

Higher harmonic currents are permitted for stage 2, provided there is a higher ratio of short-circuit power S''_k to equipment power S_G. A distinction is also made between asymmetrically-loaded equipment and symmetrically-loaded equipment. Table 2.5 gives an overview of the proposed emitted interference limits according to IEC 1000 3–4 for stage 2 equipment. The permissible relative harmonic currents for deviating ratios S''_k/S_G are determined by linear inter-polation. Limits for the total harmonic distortion (*THD*) and the partial weighted harmonic distortion (*PWHD*) are also given.

If the connection according to stage 2 is not possible, the power supply company can also allow individual exceptions according to stage 3. In this case the total harmonic current emitted from the customer system is considered in relation to the ordered power as an assessment criterion. Proposed limits of emitted interference are given in Table 2.6.

The values given in Tables 2.4 to 2.6 are presently under discussion in standards committees.

Table 2.4   Limits of emitted interference according to IEC 10003–4

Order	I_{hmax}/I_1 in %
3	21.6
5	10.7
7	7.2
9	3.8
11	3.1
13	2.0
15	0.7
17	1.2
19	1.1
21	≤ 0.6
23	0.9
25	0.8
27	≤ 0.6
29	0.7
31	0.7
≥ 33	≤ 0.6
Even-numbered	$\leq 8/h$ or ≤ 0.6

Equipment with a fundamental component of current $I_1 > 16$ A
Stage 1: $S''_k/S_G > 33$

Table 2.5   Emitted interference limits according to IEC 10003–4

Single-phase, two-phase and non-symmetrically loaded three-phase equipment

S''_k/S_G	Order h							
	3	5	7	9	11	13	Even-num-bered	*THD and PWHD*
	I_{hmax}/I_1 in %							
66	23	11	9	5	4	3	16/h	25
120	25	12	10	7	6	5	16/h	29
175	29	16	11	8	7	6	16/h	33
250	34	18	12	10	8	7	16/h	39
350	40	24	15	12	9	8	16/h	46
≤450	40	30	20	14	12	10	16/h	51

Symmetrical three-phase equipment

S''_k/S_G	Order h							
	5	7	11	13	Even-num-bered	Inter-har-monic	*THD*	*PWHD*
	I_{hmax}/I_1 in %							
66	12	10	9	6	16/h	9/h	16	20
120	15	12	12	8	16/h	9/h	18	29
175	20	14	12	8	16/h	9/h	25	33
250	30	18	13	8	16/h	9/h	35	39
350	40	25	15	10	16/h	9/h	48	46
450	50	35	20	15	16/h	9/h	58	51
≥ 600	60	40	25	18	16/h	9/h	70	57

Equipment with a fundamental component of current $I_1 > 16$ A; Stage 2

2.5.3 Compatibility levels

At the frequency-dependent impedances of the equipment, the current harmonics cause voltage drops which are superimposed on the fundamental component of voltage of the network. This, in turn, causes harmonic currents (which may be considerable, depending on the impedance of the equipment) to flow in equipment which itself generates no harmonics (e.g. capacitors). Harmonic voltages must be limited for this reason.

Table 2.7 is a summary of the specified compatibility levels of voltages for public and industrial power supply networks. The valid compatibility level for public networks is specified in EN 61000 Part 2–2 (VDE 0839 Part 2–2:1994–05)

Table 2.6 Emitted interference limits according to IEC 10003–4

Order	I_{hmax}/I_1 in %
3	19.0
5	9.5
7	6.5
9	3.8
11	3.1
13	2.0
15	0.7
17	1.2
19	1.1
21	≤ 0.6
23	0.9
25	0.8
27	≤ 0.6
29	0.7
31	0.7
≥ 33	≤ 0.6
Even-numbered	≤ 4/h or ≤ 0.6

Equipment with fundamental components of current $I_1 > 16$ A
Stage 3

for the low voltage range and in part 88:1994–03 for the medium voltage range.

The compatibility levels are identical up to the 25th harmonic in low voltage and medium voltage networks.

Compatibility levels which deviate from the values defined for public networks sometimes apply for industrial systems. They are defined in EN 61000 Part 2–4 (VDE 0839 Part 2–4:1993–06). A distinction is made between three so-called environmental classes.

Class 1: Protected supplies such as computer equipment, automation equipment, equipment for technical laboratories and protective devices.

Class 2: PCC with the public network, compatibility levels according to EN 61000–2–2 (VDE 0839 Part 2–2) and EN 61000–2–12 (VDE 0839 Part 88).

Class 3: System-internal connection points such as to welding machines, for frequent motor starting, at converter systems etc.

EN 61000–2–4 (VDE 0839 Part 2–4: 1993–06) still primarily stipulates values for the voltages at interharmonic frequencies for the industrial power supply area. For classes 1 and 2 for all interharmonic frequencies, these amount uniformly to 0.2% of the fundamental component of voltage. The values for class 3 are stipulated in Table 2.8 relative to the frequency f_{int}.

The values given in Tables 2.7 and 2.8 are permanently permissible for

Table 2.7 *Compatibility levels for harmonic voltages according to EN 61000 (VDE 0839)*

Harmonic h	Compatibility level in %				
	Low voltage systems	Medium voltage systems	Industrial systems		
			Class 1	Class 2	Class 3
Odd numbered order h, not divisible by three					
5	6.0	6.0	3.0	6.0	8.0
7	5.0	5.0	3.0	5.0	7.0
11	3.5	3.5	3.0	3.5	5.0
13	3.0	3.0	3.0	3.0	4.5
17	2.0	2.0	2.0	2.0	4.0
19	1.5	1.5	1.5	1.5	4.0
23	1.5	1.5	1.5	1.5	3.5
25	1.5	1.5	1.5	1.5	3.5
> 25	$0.2 + 0.5 \times 25/h$	$0.2 + 1.3 \times 25/h$	$0.2 + 2.5/h$	$0.2 + 12.5/h$	$5 + \sqrt{11/h}$
Odd numbered order h, divisible by three					
3	5.0	5.0	3.0	5.0	6.0
9	1.5	1.5	1.5	1.5	2.5
15	0.3	0.3	0.3	0.3	2.0
21	0.2	0.2	0.2	0.2	1.75
> 21	0.2	0.2	0.2	0.2	1.0
Even numbered order h					
2	2.0	2.0	2.0	2.0	3.0
4	1.0	1.0	1.0	1.0	1.5
6	0.5	0.5	0.5	0.5	1.0
8	0.5	0.5	0.5	0.5	1.0
10	0.5	0.5	0.5	0.5	1.0
> 10	0.2	0.2	0.2	0.2	1.0

Notes:

1) Public low voltage network EN 61000–2–2 (VDE 0839 Part 2–2:1994–05)
   Public medium voltage network EN 61000–2–12 (VDE 0839 Part 88:1994–03)
   Industrial systems EN 61000–2–4 (VDE 0839 Part 2–4: 1993–06)

2) Values for the third and ninth harmonics apply in the medium voltage range only in a.c. networks. In three-phase networks, one third of the aforementioned values should be used as the compatibility level. The compatibility levels stated apply in low voltage networks.

*Table 2.8   Compatibility levels for interharmonic
voltages according to EN 61000–2–4 (VDE 0839 Part
2–4:1993–06) for industrial power supply class 3*

Frequency f_{int} in Hz	U_{int}/U_1 in %
< 550	2.5
> 550 . . . 650	2.25
> 650 . . . 950	2.0
> 950 . . . 1150	1.75
> 1150 . . . 1250	1.5
> 1250	1.0

industrial systems. 1.5 times the values is permissible short-term, i.e. for the
duration of 10% of an interval of 150 s.

2.5.4 Interference immunity levels

As explained in Chapter 1, compatibility levels describe a value for which the
electromagnetic compatibility is obtained with a certain probability. However, it
cannot be precluded that the compatibility levels can be exceeded with respect
to either time or location. There is thus a specific probability that the
electromagnetic compatibility is not guaranteed.

The interference immunity levels of equipment must therefore be above the
particular compatibility level. General statements on this are to be found in
VDE specifications, such as in E DIN EN 50178 (VDE 0160:1998–04) (equip-
ping of power systems with electronic devices) and details on measuring and
assessing in EN 61000–4–7 (VDE 0847 part 4–7).

Product-specific standards also state interference immunity levels or permissible limits for harmonics and interharmonics or the resulting r.m.s. values.
Further details of these are not given, but some of the relevant standards and
specifications are mentioned in section 2.4 (Effects of harmonics and
interharmonics).

2.6 Examples of measurement and calculation

2.6.1 Harmonic resonance due to reactive power compensation

In an industrial network (as shown in Figure 2.29), a capacitor bank Q_C is to be
installed for reactive power compensation, so that the displacement factor
$\cos \varphi = 0.94$ is reached at the 6 kV busbar.

A general calculation equation for the resulting harmonic impedance Z_{resh} of
the resonant circuit considered from the connecting point of the converter is set
up.

On the basis of the resonant condition, the capacitor ratings Q_{Cres} (funda-

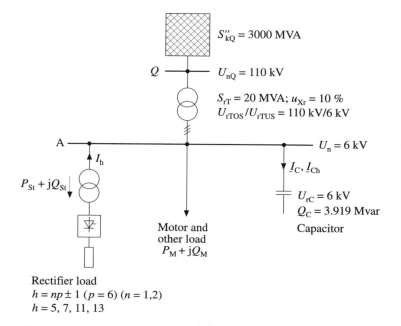

$S''_{kQ} = 3000$ MVA

Q ——•—— $U_{nQ} = 110$ kV

$S_{rT} = 20$ MVA; $u_{Xr} = 10$ %
$U_{rTOS}/U_{rTUS} = 110$ kV/6 kV

A ——•———————————————•—— $U_n = 6$ kV

I_h

$\underline{I}_C, \underline{I}_{Ch}$

$P_{St} + jQ_{St}$

$U_{rC} = 6$ kV
$Q_C = 3.919$ Mvar
Capacitor

Motor and
other load
$P_M + jQ_M$

Rectifier load
$h = np \pm 1$ $(p = 6)$ $(n = 1,2)$
$h = 5, 7, 11, 13$

Figure 2.29 *Power system diagram of a 6 kV industrial system with reactive power*
compensation
rectifier load $\underline{S}_{st} = (6 + j3)$ MVA
motor load $\underline{S}_m = (8 + j6)$ MVA

mental component powers) at which resonances $h = 5, 7, 11, 13$ occur are to be
determined and a decision is to be made as to whether the stated capacitor rating
Q_C is permissible if the $Q_{Cv} = 0.9$ Q_{Cres} to 1.1 Q_{Cres} range is forbidden.

The current harmonics fed in from the converter is calculated $I_h = f_h \times$
$(1/h) \times I_1$.

Where: $f_5 = 0.92; f_7 = 0.83; f_{11} = 0.62; f_{13} = 0.50$

The harmonic voltages U_{Ch} at the busbar and the harmonic currents I_{Ch} in the
capacitor bank are calculated.

The r.m.s. value I_C of the capacitor current is calculated and whether or not
the capacitor bank can be switched-in is assessed.

Solution

$$Z_{resh} = \frac{jX_{Nh}(-jX_{Ch})}{jX_{Nh} - jX_{Ch}},$$

because the impedance of the network supply X_{Nh} and of the capacitor X_{Ch} are
parallel from the connecting point A of the converter. Conversion produces the
following:

$$Z_{resh} = j\frac{hX_{N1}}{1 - h^2(X_{N1}/X_{C1})}$$

The impedance X_{N1} at connecting point A is

$$X_{N1} = 1.1\ U^2_n/S''_{kA} \text{ where } S''_{kA} = 204.97 \text{ MVA}$$

This produces the capacitor ratings Q_{Cres}, for which the resonance can occur, as follows:

h	5	7	11	13	
Q_{Cres}	7.543	3.803	1.54	1.103	Mvar

The forbidden area for the capacitor rating $Q_C = 3.919$ Mvar can be seen for the harmonic of the order $h = 7$, as follows:

$$Q_{Cv} = (0.9 \text{ to } 1.1)\ 3.803 \text{ Mvar}$$

$$Q_{Cv} = 3.423 \text{ Mvar to } 4.183 \text{ Mvar}$$

The currents I_h of the converter are:

h	5	7	11	13	
I_h	118.7	76.5	36.4	24.8	A

This results in the voltage harmonics at connecting point A as:

$$U_{Ch} = I_h\ Z_{resh,}$$

$$U_{Ch} = I_h\frac{X_{C1}}{h - (1/h)(S''_{kA}/1.1Q_C)},$$

whereby $X_{C1} = U_n^2/Q_C = 9.186\ \Omega$

h	5	7	11	13	
U_{Ch}	241.8	3385.5	50.1	24.2	V

The current harmonics I_{Ch} of the capacitor bank are thus

$$I_{Ch} = U_{Ch}/X_{Ch} = U_{Ch}\ h/X_{C1}$$

h	5	7	11	13	
I_{Ch}	131.6	2579.8	60.1	34.5	A

The total r.m.s. value of capacitor current is

$$I_C = \sqrt{\sum_{h=1}^{n} I_{Ch}^2};$$

with the fundamental component r.m.s. value

$$I_{C1} = Q_C/\sqrt{3}U_n = 377 \text{ A}$$

we get

$$I_C = 2611.4 \text{ A}.$$

Because the total r.m.s. value of the capacitor current is almost seven times as great as the fundamental component r.m.s. value, the capacitor bank may not be switched in.

2.6.2 Assessment of a harmonic generator

The power system shown in Figure 2.30 is assumed. A powerful twelve-pulse converter is to be connected to a 10 kV network with a finite short-circuit power. The permissibility of the connection is to be assessed on the basis of the measured current harmonics.

The following harmonic currents at rated operation were determined as measured values (95% frequency values during the assessment time period):

$$I_5 = 1.39 \text{ A}: \quad I_7 = 0.96 \text{ A}; \quad I_{11} = 14.08 \text{ A}; \quad I_{13} = 9.26 \text{ A};$$

$$I_{17} = 1.29 \text{ A}; \quad I_{19} = 0.99 \text{ A}; \quad I_{23} = 2.36 \text{ A}; \quad I_{25} = 2.63 \text{ A};$$

The impedance values Z_h of the supply would be calculated for the frequencies of the measured harmonics.

What is the magnitude of the system connection factor k_A if the total load of the industrial operation is $S_{Ind} = 6.1$ MVA?

The permissible harmonic currents I_{maxh} of the converter would be determined for the system level factor $k_{NMS} = 0.55$ and the co-phasal factor $k_{ph} = 1$ for all harmonic currents.

A decision would be made as to whether unrestricted operation of the converter is permissible.

What would the minimal short-circuit power at the point of common coupling have to be for unrestricted operation of the system to be permissible?

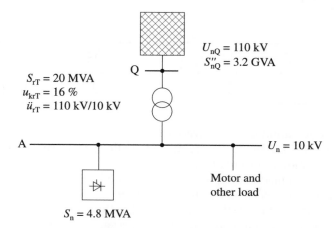

Figure 2.30   Connection of rectifier to 10 kV system (S = 4.8 MVA)

Solution

The impedance Z_h of the supply is inductive:

$$Z_h = h(X_T + X_Q)$$

with the impedance of the transformer X_T and the impedance of the 110 kV network X_Q (numerical values in %/MVA).

h	5	7	11	13	17	19	23	25	
Z_h	2.17	3.04	4.78	5.65	7.39	8.25	9.99	10.86	%/MVA

The following applies for the system connection factor k_A of the whole industrial installation.

$$k_A = \frac{S_{Ind}}{S_{rT}} = 0.153$$

The maximum permissible harmonic currents I_{hmax} are:

$$I_{h\,max} = \frac{k_A k_N U_{hmax}}{k_{ph} Z_h}$$

where

$$U_{hmax} = u_{hVT} U_n / \sqrt{3}$$

and

u_{hVT} is the compatibility level according to EN 61000–2–12 (VDE 0839 Part 88:1994–03).

h	5	7	11	13	17	19	23	25	
I_{hmax}	13.38	7.96	3.55	2.57	1.31	0.88	0.73	0.67	A

Unrestricted operation of the converter is not possible because the currents in the orders $h = 11; 13; 19; 23; 25$ are up to four times as large as the maximum permissible harmonic currents I_{hmax}.

It is therefore immediately apparent that the short-circuit power must be four times as great as the specified short-circuit power S''_k.

If it is assumed that no further consumer other than the industrial installation is connected at connecting point A, a higher system connection factor $k_A \approx 1.0$ must be used. This would make unrestricted operation of the system possible because the system connection factor is more than four times as great and the maximum permissible current is proportional to the system connection factor.

2.6.3 Impedance calculation in a medium voltage network

In the 110/30 kV network shown in Figure 2.31, a non-linear load (twelve-pulse converter to supply an industrial furnace) is to be connected at node B2. The frequency-dependent impedances were calculated to estimate the expected voltage harmonics [11]. The following system configurations of the 30 kV network are possible [12]:

- meshed network
- 30 kV cable between B2 and B3 switched off at B2
- 30 kV cable between B2 and B3 switched off at B3
- 110/30 kV transformer T103 switched off at B2
- 110/30 kV transformer T124 switched off at B3

As a significant result with regard to harmonic investigations, the frequency-dependent impedances of the B2 node in Figure 2.32 are shown for the system configurations:

- meshed 30 kV network and
- 30 kV cable between B2 and B3 switched off at B3

A series resonant point results at $f_{resR} \approx 750$ Hz and parallel resonant points at $f_{resP1} = 650$ Hz and $f_{resP2} = 850$ Hz. These resonances occur due to the parallel

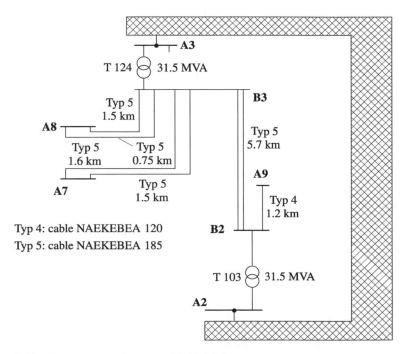

Figure 2.31   Power system diagram of 110/30 kV system

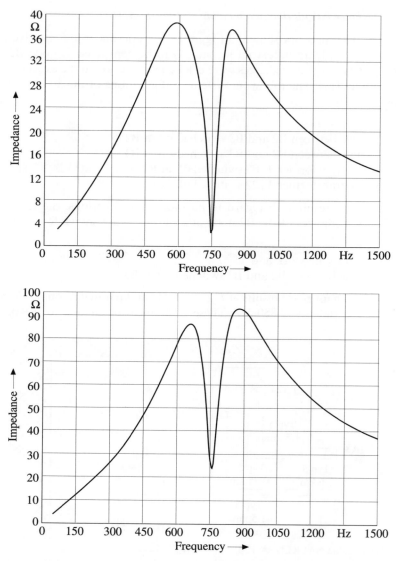

Figure 2.32   Impedance versus frequency at 30 kV busbar B2 (see also Figure 2.31)
  a) meshed system configuration
  b) 30 kV cable switched off at B3

circuit of the capacitance of the 30 kV cable B2-B3 with the inductances of the 110/30 kV transformers, where these, in turn, are to be regarded as in series with a parallel circuit of the capacitances of the 110 kV network and the inductance of the supply network.

The series resonant points of the 30 kV network node B2 remain more or less maintained for the second operational state shown (30 kV cable between B2 and B3 switched off at B3). The impedance of the parallel resonance would, of course, be substantially greater and would lead to a strong rise in the voltage harmonics for this operating state.

2.6.4 Typical harmonic spectra of low voltage consumers

The current courses and current harmonic spectra illustrated in the following (Figures 2.33 to Figure 2.37) were recorded in the electrical power generation and distribution laboratory of the University of Applied Sciences of Bielefeld using a harmonics measuring and analysis system [5,13]. These are hard copy pictures of the screen. In addition to the time course of the current, the associated harmonic spectrum of the current and the following brief information is also given.

$I(1)$ is the fundamental component r.m.s. value.
$I(\text{eff})$ is the total r.m.s. value.
$k(i)$ is the harmonic content.
THD is the total harmonic distortion.
$f(1)$ is the fundamental frequency.

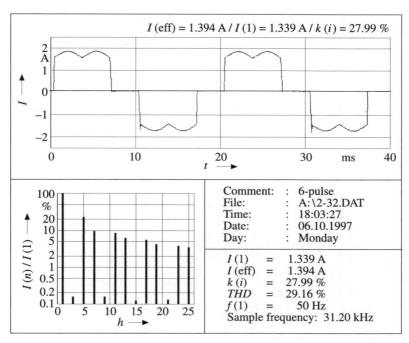

Figure 2.33 Current harmonics of six-pulse diode rectifier; $U_n = 400V$; Load $310\ \Omega$

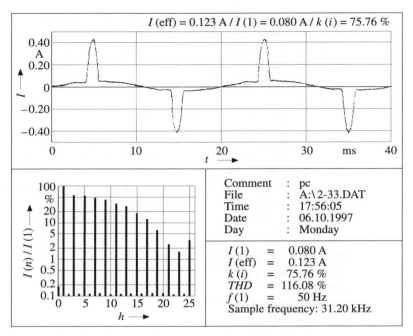

*Figure 2.34   Current harmonics of a.c./d.c. converter with capacitor smoothing
$U_n = 230$ V; $P = 130$ VA*

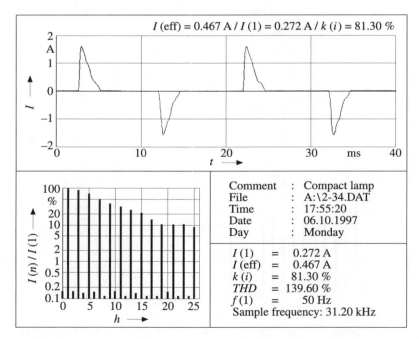

Figure 2.35   Current harmonics of compact fluorescent lamp; $U_n = 230V$; $P = 3 \times 20W$

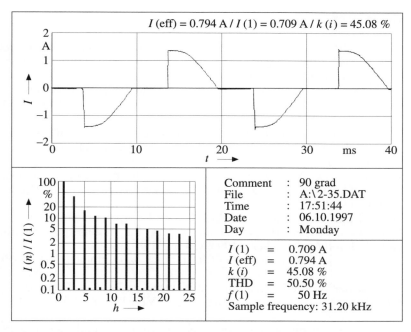

Figure 2.36   Current harmonics of dimmer; $U_n = 230V$; $P = 200W$

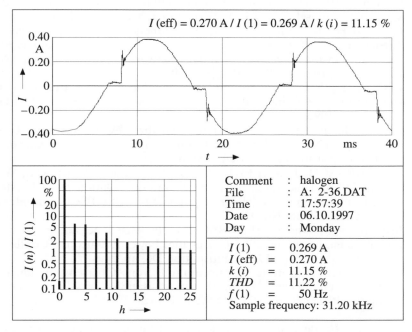

Figure 2.37   Current harmonics of electronic converter for halogen lamps, 230V, 60W

2.7 References

1   BRAUNER, G., and WIMMER, K.: 'Impact of consumer electronics in bulk areas.' Proceedings of 3rd European Power Quality Conference, Bremen, November 7–9 1995, pp. 105–116

2   BÜCHNER, P.: 'Stromrichter-Netzrückwirkungen und ihre Beherrschung (Converter system perturbations and their control)' VEB Deutscher Verlag, Leipzig (1982)

3   JÖTTEN, R.: 'Leistungselektronik. Stromrichter und Schaltungstechnik (Power electronics, converters and circuit technology)' Vieweg-Verlag, Wiesbaden (1977)

4   KLOSS, A.: 'Oberschwingungen. Beeinflussungsprobleme der Leistungselektronik (Interference problems of power electronics)' VDE-VERLAG, Berlin and Offenbach (1989)

5   MICHELS, M., and SCHLABBACH, J.: 'Experimentiersystem für Netzrückwirkungen (Experimental system for system perturbations)' *Rubrik Ausbildung und Beruf* 1994, v. 4, pp. 198–199

6   IEEE TASK FORCE: 'Effects of harmonics on equipment.' IEEE-PD 8 April 1993 v. 2.

7   'Grundsätze für die Beurteilung von Netzrückwirkungen (Basic principles for assessment of system perturbations)' VDEW, Frankfurt, 1992, 3rd edition.

8   FGH: 'Frequenzabhängige Verbraucherstrukteren und deren Zusammenwirken mit dem elektrischen Versorgungsnetz (Frequency-dependent consumer structures and their interaction with the electrical power supply system)' (Technischer Bericht 1–273 of FGH, Mannheim, 1990)

9   GRETSCH, R., and WEBER, R.: 'Oberschwingungsmessungen in Nieder- und Mittelspannungsnetzen—Netzimpedanzen (Measurement of harmonics in low and medium voltage networks—network impedances)' *Elektrizitätswirtschaft 88*, 1989, pp. 745 ff.

10   GRETSCH, R.: 'Normung und Empfehlungen (Standardisation and recommendations)' in FGH-Bericht 284 'Netzrückwirkungen' (FGH, Mannheim, 1995)

11   FGH: Referenz zu FGH-Programm: Netzoberschwingungs- und Rundsteueranalyse (NORA) [Reference to GFH program: Network harmonics and telecontrol analysis (NORA)]. Mannheim, 1997

12   SCHLABBACH, J., SEIFERT, G., WEBER, T., and WELLßOW, W.H.: 'Simulation and measurement of harmonic propagation in MV-systems – Case studies and modelling requirements. 13th PSCC, Trondheim, Norway, 1999, Paper No. 92

13   MICHELS, M., and SCHLABBACH, J.: 'PC-based measuring system for power system harmonics' 28th Universities Power Engineering Conference (UPEC) Stafford, UK, 1993, contribution D6, pp. 869–872

14   GÖKE, TH.: 'Zentrale Kompensation von Oberschwingungen in Mittelspannungsnetzen (Central compensation of harmonics in medium voltage networks)' PHD thesis, University of Dortmund, October 1997

15   GÖKE, TH.: 'Berechnung von Netzen mit Bezug auf Oberschwingungen (Calculation of networks with regard to harmonics)' in

'Spannungsqualität—Voltage Quality. Schriften aus Lehre und Forschung No. 11' SCHLABBACH, J.; ET AL.: FH Bielefeld

16   'Empfehlungen zur vermeidung unzulässiger Rückwirkungen von Tonfrequenzrundsteuerung (Recommendation for the avoidance of impermissible perturbations on audio-freqency telecontrol). 3rd revised edition, Frankfurt, VDEW, 1997

Chapter 3

Voltage fluctuations and flicker

3.1 Definitions

Changes in the amplitude of a voltage for a period which is longer than the period of the voltage under consideration is designated a voltage fluctuation.

Such voltage fluctuations can occur once, several times, randomly or regularly.

The voltage fluctuation in the form of a jump, a ramp or any quasi course is described by the value of the relative voltage change d. Figure 3.1 shows some forms of voltage fluctuations.

If voltage fluctuations occur with the frequency of approximately 0.005 Hz to 35 Hz, this generally leads, according to the amplitude, to a light flicker which can be perceived by the human eye. This subjective impression of luminance fluctuations is known as flicker. Its intensity depends on the level of the voltage fluctuation, on the frequency with which the voltage fluctuation occurs and on the type of lamp. In addition to these physical influence factors, the perception

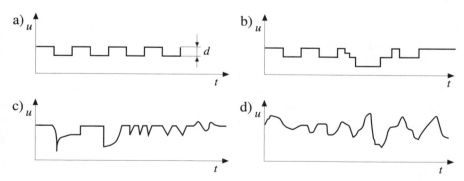

Figure 3.1   Voltage fluctuation
 a)   rectangular voltage variations
 b)   temporary irregular sequence of voltage variations with constant period
 c)   sequence of voltage variations
 d)   stochastic or steady voltage fluctuations

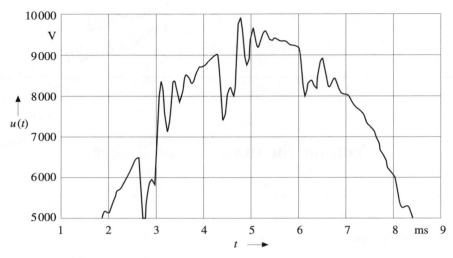

Figure 3.2   Voltage curve with commutation notches

of the flicker is also determined by the environmental conditions, as well as by the physical and psychic state of the person exposed to the flicker phenomena.

A different form of voltage fluctuation is present if the instantaneous value of the voltage characteristic deviates from the anticipated sine-wave form. Such fluctuations occur due to transient overvoltages and commutation actions. Figure 3.2 shows the voltage characteristic characterised by commutation notches.

3.2 Occurrence and causes

3.2.1 Voltage fluctuations

Voltage fluctuations can be attributed to various causes. Those voltage fluctuations caused by changes in the load situation at a system node or connection point are of interest with regard to system perturbations. The load situation at a connection point is determined by the actual composition of the individual loads but a load can also change its actual power consumption depending on the operation. Significant voltage fluctuations may be caused by the following loads:

− pulsed power output where there is burst-firing control,
− resistance welders,
− start-up of drives,
− pulsed power output with thermostat controls,
− drives with steeply-changing loading,
− arc furnaces.

Voltage fluctuations and voltage sags may occur due to system faults such as earth-leakage faults, earth short-circuits and short-circuits in the electrical power supply system. These faults impair the voltage quality at a connection point, depending on the fault location. The fault location can be in the power supply company system, in the internal system or in the system of a different power supply company which is in the 'electrical vicinity'.

Connecting and disconnecting large rectifier systems and reactive compensation systems which are controlled relative to load or reactive power can lead to voltage fluctuations. In addition to these load-related causes of voltage fluctuations, switching measures during the operation of the supply system may also lead to changes in the voltage level. Connecting and disconnecting lines can lead to voltage changes due to changes in the short-circuit power conditions, which may also be accompanied by transient overvoltages because of the actual switching operations. Fault situations in the system also lead to voltage fluctuations, or even to voltage interruptions whose strength may vary, depending on the proximity to the fault location. Although the voltage fluctuations not caused by the loads impair the voltage quality they should not be regarded as system perturbations. In a manner similar to loads, in-plant generator systems can also cause voltage fluctuations.

From a technical point of view, the voltage fluctuation results from the change to the total of all cases of reduced voltage through the impedances between the connection point and the supply sources, depending on current changes at the connection point.

3.2.2 Flicker

For voltage fluctuations which lead to flicker phenomena, a numerical assessment of the flicker level is derived from the perception of the flicker phenomena. To do this, the light source is considered to be a coiled-coil lamp (230 V, 60 W). The voltage fluctuations lead to a momentary flicker impression p_f caused by the transmission of the luminous fluctuations along the "lamp-eye-brain' chain.

The causes of flicker are the same as the causes of voltage fluctuations. Of course when considering the flicker, the physical variable, i.e. voltage, is not directly assessed, but instead the assessment is made by taking account of a special transmission function and a statistical observation over a defined time range.

3.3 Flicker calculation in accordance with empirical formulae

3.3.1 General

The flicker assessment is based on human perception of voltage fluctuations with certain outward forms and various frequencies or repetition rates. The assessment assumes that a quite special lamp is used. This lamp is a coiled-coil lamp (60 W, 230 V). Personal tests were used for various repetition rates and voltage fluctuations to determine whether a fluctuation in light could be classified from 'not visible' through 'very visible' to 'unbearable' [1].

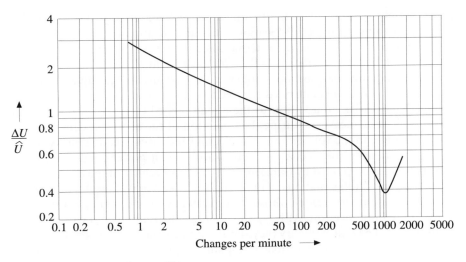

Figure 3.3   CENELEC curve [2]
Limit of disturbance for $P_{st} = 1$ for rectangular voltage variations

Figure 3.3 shows the result of these tests (CENELEC curve). Large parts of this curve can be described by simple approximation formulas.

The main influence variables in this case are the relative voltage change $d(t)$ and the repetition rate r. The shape of the curve of the voltage change is taken into account by adding form factors.

A basic prerequisite for the calculation of flicker is the determination of the relative voltage change. Where it cannot be measured it must be calculated from the supply and load data.

3.3.2 Calculation of the voltage drop in general form

When calculating the voltage drop which occurs at the connection point of a load, the procedure varies depending on the connection point of the consumer. The simpler calculation is made by considering symmetrical three-phase consumers. An equivalent circuit (as shown in Figure 3.4) can be chosen for this calculation. The impedance of the system, as well as the load, are characterised by an ohmic and inductive component.

The voltage drop using the system impedance Z_N, consisting of an ohmic and inductive component R_N and X_N, is calculated as follows:

$$\Delta \underline{U} = \Delta \underline{I}_L \cdot Z_N \tag{3.1}$$

or

$$\Delta U \cong \Delta U_R + \Delta U_X = (R \cdot \cos \varphi + X \cdot \sin \varphi) \Delta I_L \tag{3.2}$$

The short-circuit power at the connection point is determined as:

$$S_{k3}'' = \sqrt{3} \cdot U_n \cdot I_k \tag{3.3}$$

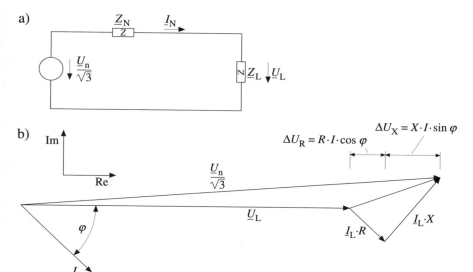

Figure 3.4 Voltage drop of symmetrical load
  a) equivalent circuit diagram
  b) vector diagram

or

$$S''_{k3} = U_n^2 / Z_N \tag{3.4}$$

The load current I_L can also be shown by the connected load S_A, as follows:

$$\Delta I_L \approx \Delta S_A / (\sqrt{3} \cdot U_n) \tag{3.5}$$

This equation is valid assuming that the voltage drop using the system impedance is low compared to the system voltage U_n.

From Equations (3.1) and (3.5) we therefore get the following approximation of the voltage drop

$$\Delta U = U_n \cdot \Delta S_A / (\sqrt{3} \cdot S''_{k3}) \tag{3.6}$$

If the a.c. current load drop (two-phase load) of the three-phase system is considered, the calculation is somewhat different. The equivalent circuit for this case is shown in Figure 3.5.

The short-circuit power at the connection point A is calculated as

$$S''_{k3} = \sqrt{3} \cdot U_n \cdot I_k \tag{3.7}$$

The change to the connected load of the consumer is

$$\Delta S_A = \Delta I_L \cdot U_n \tag{3.8}$$

assuming that the voltage drop using the system impedance is relatively low. The following applies for the voltage drop.

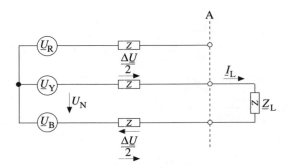

Figure 3.5   Equivalent circuit diagram of an alternating current load in a three-phase power supply system

$$\Delta U = \Delta I_{\text{L}} \cdot 2 \cdot Z_{\text{N}} \tag{3.9}$$

Equation (3.9) can be converted using Equations (3.4) and (3.8) as follows:

$$\Delta U = 2 \cdot (\Delta S_{\text{A}}/S_{\text{k3}}'') \cdot U_{\text{n}} \tag{3.10}$$

This physical voltage change is the voltage change of the phase conductor. Figure 3.6 is used to determine the voltage change of the phase-to-earth voltage.

$$\Delta U_{\text{B}} = (\sin 60°) \cdot \Delta U/2 \tag{3.11}$$

or

$$\Delta U_{\text{B}} = (\sqrt{3}/4) \cdot \Delta U \tag{3.12}$$

With Equation (3.10) we get:

$$\Delta U_{\text{B}} = (\sqrt{3}/2) \cdot (\Delta S_{\text{A}}/S_{\text{k}_3}'') \cdot U_{\text{n}} \tag{3.13}$$

By analogy this applies for ΔU_{Y}.

An alternative to the simple determination of the voltage change is shown in Figure 3.7.

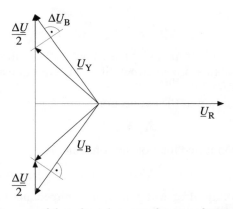

Figure 3.6   Vector diagram of the voltage between phase-conductor and earth

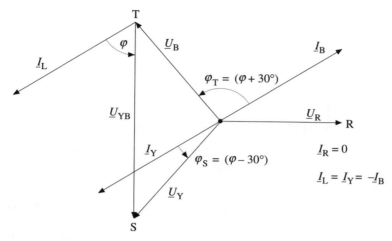

Figure 3.7   Vector diagram for alternative description (see also Figure 3.6)

Based on the knowledge of the load current change ΔI_l, the maximum voltage change ΔU of the phase-to-earth voltages can be determined as follows.
Where

$$Z_k = R + jX \qquad (3.14)$$

ΔU can be determined in accordance with the following correlation:

$$\Delta U = \text{MAX}\{\Delta U_S; \Delta U_T\} \cong \text{MAX}\{[R\cos(\varphi \pm 30°) + X\sin(\varphi \pm 30°)]\cdot\Delta I_L\} \qquad (3.15)$$

In a related representation, it follows from Equation (3.15)

$$\Delta u = \text{MAX}\{[r\cos(\varphi \pm 30°) + x\sin(\varphi \pm 30°)]\cdot\sqrt{3}\cdot\Delta S_A \qquad (3.16)$$

This corresponds to a calculation of the voltage change in accordance with the following:

$$\Delta u = \sqrt{3}(r\cdot\Delta P + x\cdot\Delta Q) \qquad (3.17)$$

The equivalent circuit shown in Figure 3.8 can be used as a basis for considering an a.c. load operated in a low voltage system between a conductor and the neutral conductor.
    The short-circuit power at the connection point A is as follows:

$$S_{k3}'' = \sqrt{3}\cdot U_n\cdot I_k \qquad (3.18)$$

or

$$S_{k3}'' = U_n^2/Z_N \qquad (3.19)$$

The load current (I_l) is determined from the supply lead voltage U_N and load impedance Z_L, as follows:

$$I_L \cong U_n/(\sqrt{3}\cdot Z_L) \qquad (3.20)$$

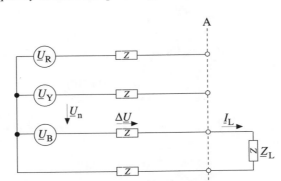

Figure 3.8   Equivalent circuit diagram of an alternating current load in a low voltage system

The connected load in this case is calculated from

$$S_A = U_n^2/(3 \cdot Z_L) \tag{3.21}$$

or

$$S_A = U_n \cdot I_L/\sqrt{3} \tag{3.22}$$

The voltage drop ΔU at the system impedance is determined from the load change ΔI_L and load impedance Z_L, as follows:

$$\Delta U = \Delta I_L \cdot Z_N \tag{3.23}$$

Using Equations (3.4) and (3.22), Equation (3.23) can also be expressed as follows:

$$\Delta U = \sqrt{3}(\Delta S_A/S_{k3}'') U_n \tag{3.24}$$

3.3.3 A_{st}/P_{st} calculation

The CENELEC curve can be approximately simulated by simplified calculations. The disturbance factor caused by a single event can be determined by the duration of the after-effect [3], as follows time duration t_f:

$$t_f = 2.3s \cdot (d \cdot F)^3 \tag{3.25}$$

where d is the relative voltage change in % and F is the form factor.

Because the individual disturbance factors are linearly superimposed, the cumulative disturbance effect is calculated from the summation of the individual disturbance factors relative to a time interval.

The flicker disturbance value for a short term interval A_{st} can be calculated as follows (time interval 10 min.):

$$A_{st} = (\sum t_f)/600 \tag{3.26}$$

The characteristic st stands for *short-term* and is generally set at 10 min. If the flicker disturbance is caused by a regular voltage change, such as determined by the repetition rate *r*, Equation (3.26) then becomes

$$A_{st} = (2.3 \ s \cdot r \cdot (d \cdot F)^3)/600 \ s \tag{3.27}$$

If the voltage fluctuations are described by a frequency, this means that the repetition rate of the voltage fluctuations is twice the value, i.e. 1 Hz corresponds to two changes per second.

For some observation time periods, the long-term flicker level is defined as that which extends over a period of two hours. To calculate this case, the value 600 s is merely replaced by the value 7200 s in Equation (3.26).

As an alternative to considering the flicker levels in the form of A_{st} values, they can also be considered in the form of P_{st} values. The approximation formula for the P_{st} value calculation is then as follows [1].

$$P_{st} = 0.36 \cdot d \cdot r^{0.31} \cdot F \tag{3.28}$$

The relationship between the A_{st} and P_{st} values are as shown in the following:

$$A_{st} = P_{st}^{\ 3} \tag{3.29a}$$

or

$$P_{st} = \sqrt[3]{A_{st}} \tag{3.29b}$$

According to Equation (3.28), the P_{st} value is proportional to the level of the voltage change. The A_{st} value on the other hand remains proportional to the repetition rate.

Section 3.3.2 details how the voltage drop calculation can be made for the a.c. or three-phase case. For the P_{st}/A_{st} calculation, the relative voltage change also can be determined using approximation formulas. This means that the relative voltage change can be directly calculated from the power change ΔS_A and the short-circuit power S_k.

The form factor *F* required for a calculation of the flicker disturbance factors A_{st} or P_{st} can, depending on the form of the voltage change, be taken from relevant graphics (Figures 3.9 to 3.11) [2].

3.4 Flicker calculation for random signals

3.4.1 Mathematical description of the flicker algorithm

The algorithm of the flicker calculation is based on the assessment of voltage fluctuations, with the simulation of the perception model of the 'lamp-eye-brain' effect chain (Figure 3.12).

The model for the lamp and the P_{st} disturbance assessment methods are described in the following sections. The transmission function for the perception of flicker phenomena is given in section 5.2.5.

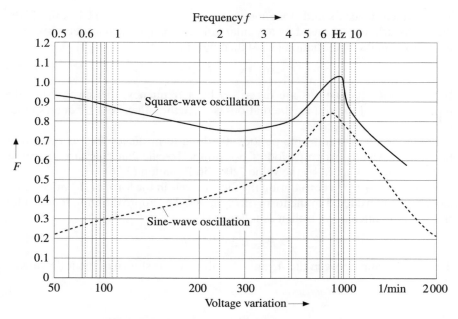

Figure 3.9   Form factor for periodical voltage fluctuations

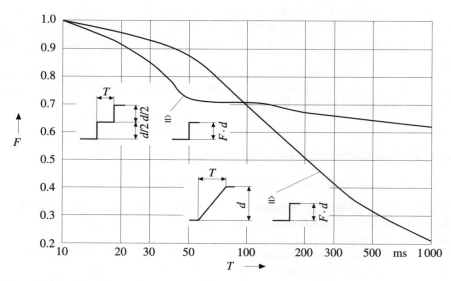

Figure 3.10   Form factor for ramps and jumps

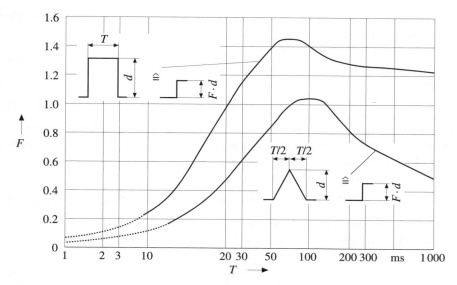

Figure 3.11   Form factor for rectangular and triangular pulses

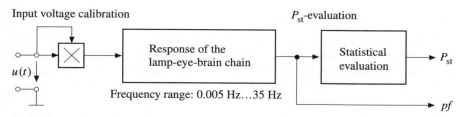

Figure 3.12   Principal structure of flicker meter

The transmission function of the coiled-coil lamp is an essential part of the flicker algorithm.

The transmission function of the light source can be simulated for the relationship between voltage fluctuations and changes in the light flux. For general-use filament lamps a tungsten coil is heated to a high temperature. The active power, $P_L(t)$ is proportional to the temperature of the coil. The light flux follows the temperature without lag. Fluctuations in the light flux are attenuated by the inertia of the coil [3].

$$P_L(t = (U^2/R)[1 + \cos(2\omega t)] \tag{3.30}$$

For small temperature fluctuations $\Delta\vartheta$, the differential equation of the filament lamp can be given as follows:

$$m_w c[\mathrm{d}(\Delta\vartheta)/\mathrm{d}t] + P_m/C_L = P_{ab}(t) \tag{3.31}$$

where

$$P_{ab}(t) = P_L(t) - P_m, \ P_m = U^2/R \tag{3.32}$$

Assuming that Φ is proportional to $\Delta\vartheta$, Equation (3.27) gives us the following for the light flux.

$$\Phi(t) = \Phi \cos (2\omega t - \varphi) \qquad (3.33)$$

where

$$\Phi(t) = K/\sqrt{1 + (2\omega\tau)^2} \qquad (3.34)$$

The transmission function between the fluctuations of the electrical power of the filament lamp and the fluctuations in the light flux correspond to a low-pass filter of the first order.

For an amplitude-modulated voltage signal

$$u = u \sin (\omega t)[1 + m \sin (\omega_f t)] \quad \text{when } m = \Delta U/U \qquad (3.35)$$

the light flux is also amplitude modulated assuming $m \ll 1$ with ω_f

The following applies for coiled-coil lamps.

$$\frac{\Delta\Phi/\Phi}{\Delta U/U} = \frac{3.8}{\sqrt{1 + (\omega_f\tau)^2}} \qquad (3.36)$$

3.4.2 The P_{st} disturbance assessment method

With the P_{st} disturbance assessment method, the momentary flicker impression is transferred to the flicker level [3]. The momentary flicker is classified over the time period of the measured interval. The relative frequency of the momentary flicker is then determined from these values. The values of the relative cumulative frequency are then determined from the values of the relative frequency. From the course of these values, the flicker level is determined by evaluating certain points. Figure 3.13 shows the possible course of the relative cumulative frequency for a measured interval. At stipulated cumulative frequency values, the level of the momentary flicker impression is evaluated with the equation of condition (3.37).

The values P_i show which momentary flicker level was exceeded for i percent of the observation time.

$$P_{st} = \sqrt{0.0314 \, P_{0.1;g} + 0.0525P_{1.0;g} + 0.0657P_{3.0;g} + 0.28 \, P_{10;g} + 0.08P_{50;g}} \qquad (3.37)$$

The assessment method is called a smoothed assessment method if the individual values P_i were determined from several supporting values.

$$P_{0.1;g} = P_{0.1}$$

$$P_{1.0;g} = (1/3)(P_{0.7} + P_{1.0} + P_{1.5})$$

$$P_{3.0;g} = (1/3)(P_{2.2} + P_{3.0} + P_{4.0})$$

$$P_{10;g} = (1/5)(P_{6.0} + P_{8.0} + P_{10} + P_{13} + P_{17})$$

$$P_{50;g} = (1/3)(P_{30} + P_{50} + P_{60}) \qquad (3.38)$$

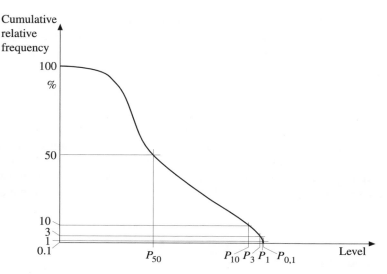

Figure 3.13 P$_{st}$-flicker evaluation

A further important variable for assessing the flicker phenomena is provided by the long-term flicker level P_{lt}:

$$P_{lt} = \sqrt[3]{(1/N)\sum_{i=1}^{N} P_{sti}^3}$$ (3.39)

For this assessment, high flicker levels are particularly highly assessed. The observation time period in this case is generally 2 h ($N = 12$). P_{lt} is determined from a sliding measuring interval.

The advantage of flicker measurement is the direct transfer of voltage fluctuations of various forms and amplitudes to an assessment number.

3.5 Effects of voltage fluctuations

Voltage fluctuations cause the disturbing effect of luminance fluctuations which are usually perceived before there is any effect on the operation of components or equipment. The voltage fluctuations also result in quite different disturbance phenomena. These include the following:

- control actions for control system acting on the voltage angle,
- braking or acceleration moments from motors connected directly to the system,
- impairment of electronic equipment where the fluctuation of the supply voltage passes through the power supply assembly to the electronic equipment.

This last point, in particular, is of great importance. Disturbance phenomena of this kind can occur in equipment for all applications. The following equipment is particularly worthy of mention:

- computers, printers, copiers,
- monitoring equipment,
- control units, control computers,
- components for telecommunication.

Voltage fluctuations due to commutation notches also lead to the effects already mentioned. They also particularly affect capacitor charging. The relative steep-edged commutation notches can also, under certain circumstances, generate resonance points of very high frequency (a few kilohertz) in electrical systems.

3.6 Standardisation

Standards are assigned to fixed application areas. In addition to the standardisation of the compatibility level for public and industrial systems, there are also standards for assessing the emitted interference and interference immunity.

The voltage quality is defined in EN 61000–2–2 (VDE 0839 part 2–2) in public low voltage systems. The valid compatibility levels for industrial systems are given in EN 61000–2–4 (VDE 0839 parts 2–4). EN 50160 stipulates the compatibility levels for public medium and low voltage systems.

The guide values for the assessment of flicker disturbance are given in Table 3.1.

The permissibility of a connection in a low, medium or high voltage system is assessed using the method shown in Table 3.2. If, when a customer's system is

Table 3.1　Guide values for the assessment of flicker disturbance [2]

	A_{lt}	A_{st}	d
Permissible disturbance factor			
Low voltage	0.4	1	
Medium voltage	0.3	0.75	
High voltage	0.2	0.5	
Permissible disturbance factor for a customer system*			
Low voltage	0.05	0.2	0.03
Medium voltage	0.05	0.2	0.02*
High voltage	0.05	0.2	0.02*

* Higher values are acceptable in exceptional cases.

Table 3.2   Assessment method for flicker level [2]

Requirement for A_{st} and A_{lt}	Consequences for the connection
$A_{st} < 0.2$ and $A_{lt} < 0.05$	Admissible
$0.2 < A_{st} < 0.5$ or $0.05 < A_{lt} < 0.2$	Qualified admissible
$A_{lt} > 0.2$	Inadmissible. Measures required

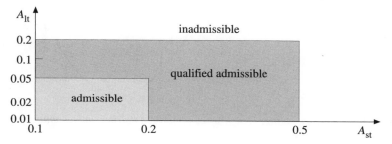

Figure 3.14   Assessment of flicker

connected, an A_{st} value of 0.2 and an A_{lt} value of 0.5 are not exceeded, a connection is essentially permissible. In special cases a higher disturbance factor can be assigned to an individual customer. This usually applies if other customers connected at a node do not use the share of the overall disturbance factor assigned to them. In this case it must be considered that one single customer is responsible for the disturbance emission in the long-term range for the maximum A_{lt} value of 0.2 (Figure 3.14).

The connection is permissible for a ratio of more than 1000 of the short-circuit power at the node to the connected load of the customer. This criterion is of secondary importance in the case of a single-phase load.

The limits for the current emission of equipment are divided into classes for equipment where $I_N \leq 16$ A and those where $I_N > 16$ A IEC 1000–3–5 resp. (VDE 0838 part 5). In addition to the limits, this standard also contains the test conditions for equipment manufacturers.

The interference immunity for equipment is specified in EN 50178 (VDE 0160). Limits in accordance with Figure 3.15 are permissible for commutation notches. The electrical equipment must continue to function properly where there is a voltage signal, as shown in Figure 3.15.

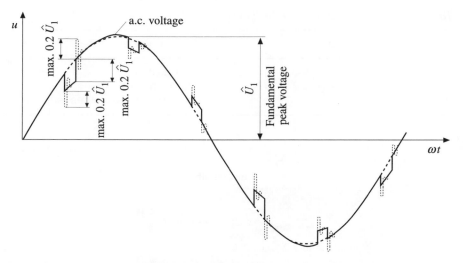

Figure 3.15   Maximum voltage change by commutation notches and commutation oscillations

3.7 Examples of measurement and calculation

3.7.1 Measurement of flicker in a low voltage system

The result of a flicker measurement in a low voltage system is shown in Figure 3.16. The measurement was taken at the 400 V voltage level of a distribution transformer ($S_N = 630$ kVA). This transformer supplies a small light industrial area. The measuring results show a pronounced daily profile. During the day the flicker level is clearly above the value of $P_{st} = 0.7$ or $P_{st} = 1$, but nevertheless no complaints whatsoever occurred in this case.

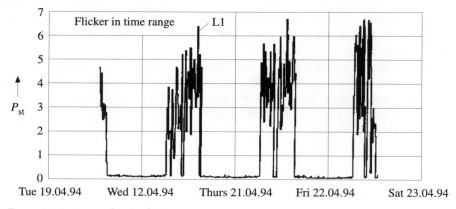

Figure 3.16   Example of a flicker measurement

3.7.2 Calculation of an industrial system for resistance heating

Let us assume resistance heating which is connected via a transformer to a 10 kV busbar.

The resistance heating is asymmetrically connected between two conductors. The busbar is fed via two parallel 1.5 km long mass-impregnated, paper-insulated cables of 185 mm² conductor cross-section. The layout is shown in Figure 3.17.

The following equivalent impedances result for the equivalent circuit.

Location Q

$$X_{kQ} \approx Z_{kQ} = U_n^2/S_{kQ}^{''} \Rightarrow X_{kQ} = 0.529 \ \Omega$$

Cable

Using a table, the following quantities per unit length were determined.

Resistance quantity: 0.164 Ω/km
Reactance quantity: 0.090 Ω/km

From this we get the following for the resulting equivalent circuit elements of the cable.

$$X_L = (1/2) \times 0.090 \ (\Omega/km) \times 1.5 \ km = 0.068 \ \Omega$$
$$R_L = (1/2) \times 0.164 \ (\Omega/\neq km) \times 1.5 \ km = 0.123 \ \Omega$$

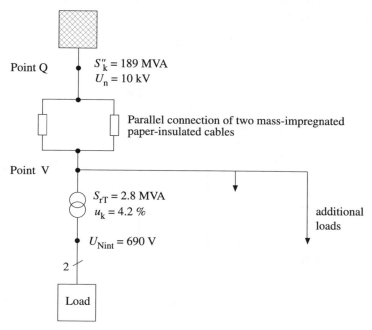

Figure 3.17   Single-line diagram of a low voltage system

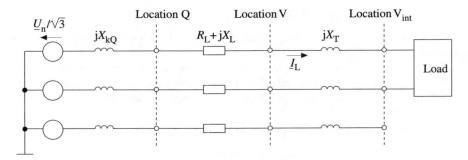

Figure 3.18   Equivalent circuit diagram (see also Figure 3.17)

Transformer (high voltage side):

$$R_T \approx 0$$

$$\Rightarrow Z_T = X_T = u_k \times (U_n^2/S_{rT}) \Rightarrow Z_T = 1.5 \; \Omega$$

From these values we get the equivalent circuit shown in Figure 3.18.

The heating represents a base load of 600 kW with regular square-wave power changes occurring around 800 kW (Figure 3.19).

The connection of the load should be assessed for flicker at location V.

Where there is a base load of 600 kW, there is a voltage drop $U_{0\Delta}$ (allowing for line-to-line voltage) of

$$U_{0\Delta} = 2 \times I_L \times Z_{tot}$$

where

$$Z_{tot} = \sqrt{(X_{kQ} + X_L)^2 + R_L^2}$$

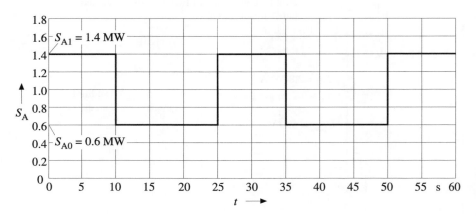

Figure 3.19   Change of apparent power

$$I_L = S_{A0}/U_n$$

$$Z_{tot} = 0.6095 \ \Omega$$

$$I_L = 60 \ A$$

$$U_{0A} = 73.14 \ V$$

If the power changes to 1.4 MW, we get the following for the voltage drop.

$$U_{1A} = U_{0A} \times (1.4 \ MW/600 \ kW) = 170.66 \ V$$

Therefore, the following applies for the voltage change.

$$\Delta U_A = U_{1A} - U_{0A} = 97.52 \ V$$

The change to the phase-to-earth voltage is:

$$\Delta U_Y = (\sqrt{3}/4) \times \Delta U_A \Rightarrow \Delta U_Y = 42.23 \ V$$

Therefore the relative voltage change is:

$$d = (\sqrt{3} \ \Delta U_Y/U_n) = 0.0073 \Rightarrow 0.73\%$$

From this it follows that further investigations are necessary.

According to the VDEW brochure [2] the following duration time of flicker after-effect t_f results:

$$t_f = 2.3 \ s \times (100 \times d \times F)^3$$

The form factor F for the voltage change course resulting from Figure 3.16 is $F = 1$.

From this we get:

$$t_f = 0.895 \ s$$

The following applies for the flicker disturbance factors A_{lt} and A_{st}:

$$A_{st} = \frac{\sum\limits_{0=0}^{n} t_f}{10 \times 60 \ s} \quad \text{(10-minute interval)}$$

$$A_{lt} = \frac{\sum\limits_{0=0}^{n} t_f}{120 \times 60 \ s} \quad \text{(2-hour interval)}$$

In the concrete case we therefore get:

$$A_{st} = \frac{50 \times t_f}{10 \times 60 \ s} = 0.0746$$

Because it is a signal which is constant over time, A_{st} and A_{lt} are identical in this case.

$$A_{lt} = \frac{600 \times t_f}{120 \times 60 \ s} = 0.0746$$

According to the VDEW brochure [2], this means that the connection is possible only in exceptional cases, because:

$$0.05 < A_{lt} < 0.2$$

On the contrary, if the frequency is reduced to 400 in 2 h, we get:

$$A_{lt \ 1} = \frac{400 \times t_f}{120 \times 60 \ s} = 0.0497$$

In this case the connection is just permissible.

3.8 References

1   NEVRIES, K.-B., and PESTKA, J.: 'Bewertung der Flickerwirkung von Spannungsschwankungen in öffentlichen Versorgungsnetzen und die abgeleitete zulässige Störemission einzelner Kundenanlagen (Assessment of the flicker effect of voltage fluctuations in public electrical supply systems and the derived permissible disturbance emission of individual customer systems)' *Elektrizitätswirtschaft 86*, 1987, pp. 245–250

2   'Grundsätze für die Beurteilung von Netzrückwirkungen (Basic principles for the assessment of system perturbations)' (Verlags- und Wirtschafts-gesellschaft der Elektrizitätswerke m.b.H (VWEW), Frankfurt, 1987, 2nd edn.)

3   MOMBAUER, W.: 'Digitale Echtzeit-Flickermeßtechnik (Digital real-time flicker measuring techniques)' FGH-Bericht (Report), 1993, pp. 1–279

Chapter 4

Voltage unbalance

4.1 Occurrence and causes

Voltage unbalances occur in electrical power supply systems due to the asymmetry of the equipment on the one hand and the asymmetry of load states on the other. The main influencing factor with regard to the equipment can be overhead lines. Because of the geometric arrangement, the different mutual influence and the different phase-to-earth capacity lead to asymmetries.

With regard to system perturbations, the asymmetrical load states cause the asymmetries. In low voltage systems they occur mainly due to the numerous a.c. loads connected between the phase-conductors and the neutral conductor. This 'normal' consumer mode is also the reason why the low voltage system is operated in the form of a low-impedance earthed (TN) system. At the medium and high voltage levels the a.c. loads operated between the two conductors form the rare loads. Some typical loads in this category are as follows:

- arc furnaces,
- resistance melting furnaces,
- traction supplies,
- heavy-current test systems.

Because of their design and operating mode, the loads which cause asymmetries create voltage fluctuations at the same time and are thus significant with regard to flicker phenomena.

4.2 Description of unbalances

4.2.1 Simplified examination

The unbalance of the voltage is defined by the relationship between the negative sequence system and positive sequence system of symmetrical components, as follows:

$$k_U = U_2/U_1 \tag{4.1}$$

k_U can be approximately determined as follows:

$$k_U \cong S_A/S''_{k3} \tag{4.2}$$

More accurate examinations require a more complicated calculation.

4.2.2 Symmetrical components

The definition of voltage unbalance is based on the representation of the three-phase system in the form of symmetrical components. According to the transformation rule, each three-phase system is represented by the super-imposition of two symmetrical three-phase systems and one a.c. system. The three-phase system consists of the positive sequence system and negative sequence system, a system which rotates counterclockwise. The a.c. system is called a zero sequence system.

The transformation of the voltages of the three-phase system was described in detail in 1.4.3 for phase-to-earth voltage. The following applies.

$$\begin{bmatrix} \underline{U_0} \\ \underline{U_1} \\ \underline{U_2} \end{bmatrix} = \frac{1}{3} \begin{bmatrix} 1 & 1 & 1 \\ 1 & \underline{a} & \underline{a}^2 \\ 1 & \underline{a}^2 & \underline{a} \end{bmatrix} \cdot \begin{bmatrix} \underline{U_R} \\ \underline{U_Y} \\ \underline{U_B} \end{bmatrix} \tag{4.3}$$

The derivations for the phase-to-earth voltages can also be used completely analogously for phase-to-phase voltages.

Both calculations result in the same degree of asymmetry with regard to amount. Figure 4.1 shows the phasor diagrams of various three-phase systems.

The illustrated symmetrical system does not result in any asymmetry. The three-phase system with a zero sequence system, but with symmetrical phase-to-phase voltages, has no asymmetry as considered here. The asymmetrical phase-to-earth voltages with symmetrical phase-to-phase voltages are generally found in medium voltage systems with earth-fault compensation using Petersen coils.

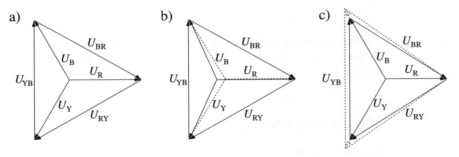

Figure 4.1   Vector diagrams of three-phase supply systems
   a) symmetrical system
   b) symmetrical system with neutral (zero sequence) system
   c) unsymmetrical system

The differences between phase-to-earth voltages purely with regard to amount are not a measure of the asymmetry. The system with the different phase-to-phase voltages has an asymmetry in the meaning of the definition.

4.3 Effects of voltage unbalance

Voltage unbalances on drive machines lead to increased losses. In the case of synchronous machines, the current of the negative sequence system should remain limited to values of 5% to 10% of the rated current. On asynchronous machines, voltage unbalances of even 2% can lead to damaging temperature rises. For power electronic circuits where the firing angle is derived from the voltages, asymmetries cause ripple in the generated d.c. voltage. In twelve-pulse circuits, asymmetry leads to a 100 Hz component of the d.c. voltage and to a harmonic current component in the order of $h = 3$ in the system current.

4.4 Standardisation

The emitted interference of an individual disturbance should not exceed a value of $k_U = 0.7\%$ in the area of the asymmetry [1]. The compatibility level for medium voltage systems is stipulated as 2%. This value is given in EN 61000–2–2 and EN 61000–2–4 (VDE 839 part 2–2 and part 2–4). EN 50160 also stipulates a value of 2%. Ten-minute mean-value intervals should be assessed to determine the level.

The interference immunity for electrical equipment is stipulated as 2% according to EN 50178 (VDE 0160). EN 50178 also stipulates an asymmetry for the ratio of the voltage of the zero phase-sequence system to the voltage of the positive sequence system. A limit of 2% is also stipulated for this quantity.

4.5 Examples of measurement and calculation

4.5.1 Measurement of unbalance in an industrial 20 kV system

Figure 4.2 shows the course of the voltage unbalance over a period of 12 days in an industrial 20 kV power supply district with a peak load of approximately 7.8 MW. The measurement is taken on the secondary side of the supply transformer. There are several decentralised supplies within the supply district.

The mean values of the voltage unbalance are recorded over 60 s in each case. The periodic of the unbalance corresponds to the load periodic in the supply district. August 18th is a statutory holiday in the supply area under consideration. The measuring period clearly exceeds the load period. Further measurements showed that the asymmetry peak on 18.08 at 16.24 hours was caused by a single-pole short interruption in the superimposed 400 kV level.

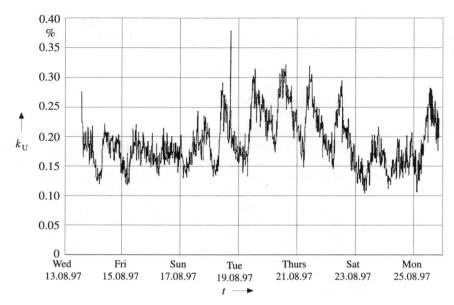

Figure 4.2   Unsymmetrical voltage component in a medium voltage system

Despite the mainly industrial consumers, the voltage asymmetry is clearly below the permissible compatibility level of 2%. The actual detected 95% level is at 0.28% for the measurement.

4.5.2 Determining the unbalance of an industrial system

For this example, the system detailed in section 3.7.2 is again used and the unbalance for location V considered.

The short-circuit power at location V is obtained from:

$$S''_{kv} = U_n^2/Z_{tot} = 160 \text{ MVA}$$

The connection of an asymmetric load is permissible at:

$$S_A/S''_{kV} < 0.7\%$$

For the basic load (S_{A0}) of 600 kW, the following applies:

$$S_{A0}/S''_{kv} = (600 \text{ kW})/(164 \text{ MVA}) = 3.7 \times 10^{-3} < 0.7\%$$

At a maximum load (S_{A1}) of 1.4 MW, the following applies:

$$S_{A1}/S''_{kv} = 8.5 \times 10^{-3} > 0.7\%$$

The emitted interference is above the emitted interference of 0.7% permissible for a single system. However the connection can sometimes be approved in individual cases.

4.6 Reference

1   'Grundsätze für die Beurteilung von Netzrückwirkungen (Basic principles for assessment of system perturbations)' (Verlags- und Wirtschaftsgesellschaft der Elektrizitätswerke mbH (VWEW), Frankfurt, 1987, 2nd edn.)

Chapter 5

Measurement and assessment of system perturbations

5.1 General

The increasing use of equipment and loads with a non-linear current–voltage characteristic and/or operating characteristics which are not steady over time, has led to an increase in system perturbations in electrical power supply systems of public power supply and industrial networks. In parallel with the development of suitable standards and recommendations for the definition of limits and compatibility levels, measuring procedures and instruments are being developed which enable the relevant measured quantities for system perturbations to be acquired. The following quantities are of particular interest:

- voltage fluctuations,
- flicker,
- transient overvoltages,
- voltage unbalance,
- harmonics,
- interharmonics.

Figure 5.1 summarises these quantities relative to the frequency range to which the measured quantities are to be assigned. The amplitudes at which the individual quantities occur are also given.

It is not possible to make a precise statement regarding the frequency range for voltage fluctuations. The amplitudes are within a range of a few percentage points of the r.m.s. value. For flicker, the frequency is in a range from a few millihertz up to approximately 35 Hz. The amplitudes are in a range up to a few percentage points.

For harmonics, the spectrum is at present generally considered up to a frequency of 2.5 kHz. The amplitudes of the voltage are also in the order of a few percentage points. For current harmonics the values can be in the magnitude of the fundamental component or even higher. Voltage unbalances are generally in

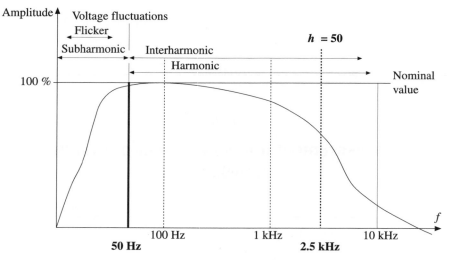

Figure 5.1 Frequency range of perturbations

the order of 1% to 2% and are relative to the fundamental component (see Chapters 2 and 3).

5.2 Sampling systems

5.2.1 General characteristics

With the introduction of digital technology, instruments operating in the time domain have been pushed into the background more and more. The technology used in the measuring instrument market has improved rapidly. Computers have become increasingly powerful, characterised by a growing number of computing operations per time unit. Digital signal processing is also constantly entering new fields with regard to sampling frequency and amplitude resolution. Despite this, their use remains cost effective.

The two quantities essential for digital signal processing are the sampling frequency and amplitude resolution.

Figure 5.2 shows how an analogue measured signal is converted to a value-continuous sampling sequence by sampling in the time range. If the amplitude is then converted to discrete amplitude values, e.g. using an analogue-digital converter, this produces a value sequence which can be processed by a computer or digital signal processor (DSP).

5.2.2 Basic structure of a digital measuring instrument

The basic structure of a digital measuring instrument consists of a few components (see Figure 5.3). The measured signal is decoupled by an input adapter.

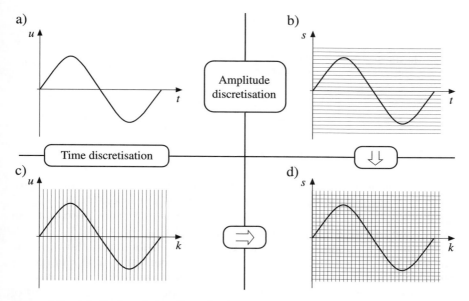

Figure 5.2   Amplitude and time discretisation
  a)   continuous in time/continuous in range
  b)   continuous in time/discrete in range
  c)   discrete in time/continuous in range
  d)   discrete in time/discrete in range

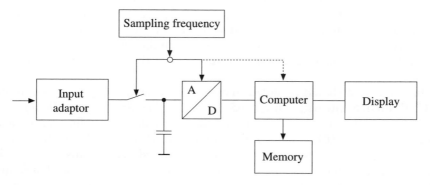

Figure 5.3   General block-structure of digital measurement systems

This assembly limits the frequency range of the measuring instrument to the working range and protects the electrically-sensitive microelectronics. The frequency band is limited by a low-pass filter, i.e. an antialiasing filter.

An A/D converter converts the continuous analogue signal to a sampling sequence which is discrete with regard to both amplitude and values. The sample and hold elements are fitted between the input adapter and converter. The purpose of this component is to keep the signal to be measured constant for the period of the A/D conversion.

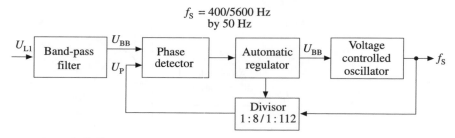

Figure 5.4   Phase-Locked-Loop unit (PLL)

The sampling sequence provided by the A/D converter is then further processed by the arithmetic logic unit. Nowadays, this is a microcontroller, a digital signal processor or a complex processor system. This arithmetic logic unit also controls the display and generally any memory units.

The sampling frequency and other control signals for the arithmetic logic unit, and also for any display, are generated by a controller. This can consist essentially of a crystal generator with corresponding divider stages or a phase-locked-loop (PLL) assembly. The generation of sampling and control frequencies using a crystal generator produces sampling sequences where the sampling impulses always occur precisely at the same intervals. This means that time sections from a recorded sampling sequence can be readily determined. This type of sampling frequency generation is used on oscilloscopes and other recording instruments, such as transient recorders. If, however, a sampling sequence is required which is in a numerically-fixed relation to a dominant frequency component in a measured signal, and if the frequency of this signal is subject to certain fluctuations, only a PLL can then be considered for generation of the sampling frequency. The structure of this component is shown in Figure 5.4.

In this case it should be noted that the frequency in the grid system of the UCTE can fluctuate within the 49.95 Hz to 50.05 Hz range [1] (see Figure 5.5).

A direct assignment of measured signals to specific time points is no longer possible where a PLL is used. Either the frequency of the PLL or the corresponding equivalent numerical value must be stored. The timing can then be reconstructed using these values. The PLL is used mainly for harmonics analysers and flicker meters.

5.2.3 Transient recorders

Recording instruments or transient recorders are used to measure voltage fluctuations. The voltage fluctuation must then be evaluated using the time course of the measured voltage. The principle of measuring instruments for recording voltage fluctuations is shown in Figure 5.6.

The characteristic features of a transient recorder are its amplitude resolution, its sampling frequency and its memory depth. The amplitude resolution is in the

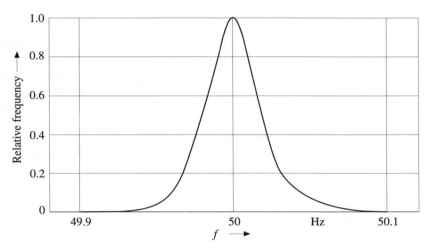

Figure 5.5   Relative frequency of UCTE network frequency
(measuring time: one year)

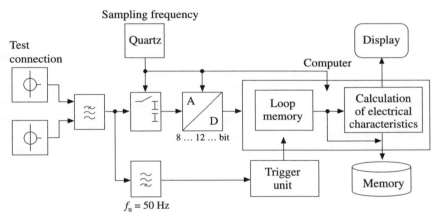

Figure 5.6   General block-structure of a transient recorder

12-bit to 14-bit range (4.096 levels up to 16.384 levels). The sampling frequencies range from approximately 10 kHz up to 100 kHz. Transient recorders with sampling frequencies of a few megahertz are now available for recording very fast signals, e.g. transient overvoltages. These instruments usually have an amplitude resolution of between 8 bits and 10 bits (256 levels up to 1.024 levels).

The recorded measured values can be printed directly. Special program packages or general statistical or tabular calculation programs can be used for further analysis of the measured values using a computer.

5.2.4 Harmonics analysers

Various kinds of measuring instruments have long been in use for measuring harmonics. Harmonics analysers can, for instance, be designed on the basis of selective filters coupled with r.m.s. value measurement. Such instruments are now rarely found in use. Because of the technical developments in computer technology, instruments consisting of sampling systems and which calculate harmonic components using Fourier transformation, or discrete Fourier transformation (DFT), are more commonly used.

A harmonics analyser which determines the harmonic components using Fourier transformation consists of the following components:

- measured signal coupling/amplifier,
- antialiasing filter,
- sample and hold elements,
- multiplexer (if required),
- A/D converter
- computer unit,
- display unit,
- storage medium,
- unit for generating the sampling frequency and a controller.

The named components are combined in principle as shown in Figure 5.7.

The measured signal can be input either galvanically separated or galvanically coupled. It is then amplified for the individual measurement ranges so that the best possible control of the A/D converter results. These components can also compensate for the fundamental component. The measured signals are then applied to the antialiasing filter and band-limited. After this initial processing, the signals are then passed to the sample and hold elements.

The assemblies described up to now are provided for each measurement

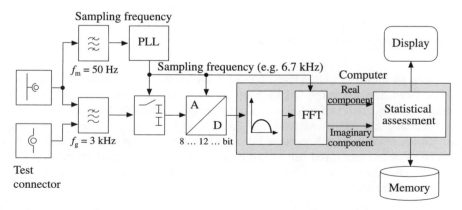

Figure 5.7   General block-structure of a measurement system for harmonics

channel. Depending on the design of the instrument, the measured signals of the individual channels are either passed via a multiplexer to a central A/D converter, or each measurement channel may have its own converter. A/D converters in use today mainly have a resolution of 12 to 16 bits (4,096 to 65,536 quantisation levels). The digitalised measured values are applied to the computer unit where they are analysed. From here, the measurement results are passed to the display, statistically processed and stored as necessary. The measuring instrument has a controller which contains as an essential component, a unit for generation of the sampling frequency. This is generally a precision timebase combined with a PLL.

5.2.5 Flicker meter

Either a transient recorder or flicker meter can be used to measure flicker levels. The flicker meter shows the instantaneous flicker impression p_f and flicker level P_{st} or A_{st} relative to an adjustable measuring interval (1 minute, 5 minutes or 10 minutes) as a direct measured value. The principle of construction of a flicker meter using digital technology is shown in Figure 5.8 [2].

The flicker meter consists of various functional blocks [3]. The first block regulates the amplification of the measured voltage. The measured voltage is corrected to 100% via a first order low-pass filter with a correction time of 60 s. This step enables the voltage changes to be considered as relative quantities.

Block 2 considers the squaring in the lamp transmission function ($\Phi = U^2$). The signal is demodulated in a low-pass filter in block 2. This is a Butterworth low-pass filter of the sixth order. At the same time it suppresses the signal components with a double modulation frequency produced by the squaring. In block 3 the low-pass characteristic of lamps is simulated and this block also shows the form filter with a band pass characteristic for simulating the transmission function of the human eye.

$$F(s) = \frac{k\omega_1 s}{s^2 + 2\lambda s\omega_1^2} + \frac{1 + \dfrac{s}{\omega_2}}{\left(1 + \dfrac{s}{\omega_3}\right)\left(1 + \dfrac{s}{\omega_4}\right)} \tag{5.1}$$

The parameters are:

k $= 1.74802$
λ $= 2\pi\, 4.05981$
$\omega_1 = 2\pi\, 9.15454$
$\omega_2 = 2\pi\, 2.27979$
$\omega_3 = 2\pi\, 1.22535$
$\omega_4 = 2\pi\, 21.9$

Block 4 contains a variance estimator which is achieved by squaring with first order low-pass filtering ($\tau = 300$ ms). The signal of the instantaneous flicker level

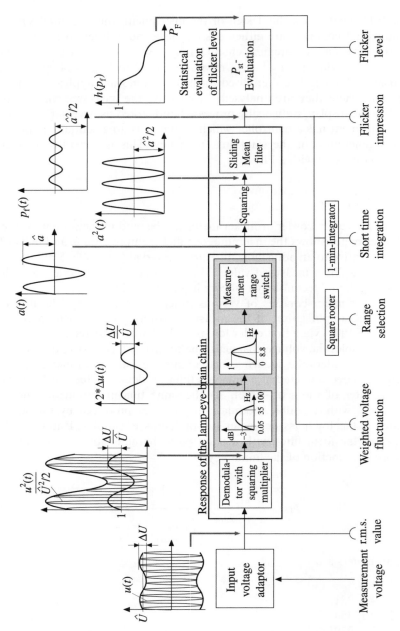

Figure 5.8   General block-structure of a flicker meter with signals

p_f is present at the output of this block. This level is statistically evaluated in block 5 using the P_{st} disturbance evaluation method (see section 3.4.2).

5.2.6 Combination instruments

Various instruments can be used for measuring and investigating the voltage quality and determining the system perturbations, depending on the individual aspects to be considered. Table 5.1 is a summary of the assignment of instruments to the individual aspects of the determination of the voltage quality by measurement.

Because of the high degree of integration that can be achieved with today's technology, measuring instruments are available that are a combination of transient recorders, harmonics analysers, flicker meters and oscilloscopes. These instruments are sometimes able to perform the individual measurement functions simultaneously. Furthermore, these instruments are fitted with special analysis functions and suitable software to evaluate the measurements.

5.3 Measured value processing

5.3.1 Statistical methods

The instantaneous values of a characteristic are not evaluated directly when assessing system perturbations. Measurement results are assessed from a statistical viewpoint relative to compatibility levels. Overshoots of compatibility levels are permissible for short time periods. This applies, for example, for disturbed operating conditions and switching operations (see also section 1.1). Further aspects for usage of measuring instruments are: suitability for field use, ease of operation, data exchange, suitability for calibration.

Table 5.1 Assignment of measuring instruments

Measuring instrument / x-t recorder	Voltage fluctuation	Flicker	Unbalance	Harmonics	Interharmonic	Measured time period	Accuracy	Complex evaluation Present	Complex evaluation Possible
mechanical	0	−				up to days	+	No	No
electronic	+	−				up to days	0 to +	No	Yes
storage oscilloscope	+	−	−			short	0 to +	No	Yes
transient recorder	+	0	+	0	0	short	0 to +		
spectrum analyser									
laboratory instruments			0	+		short	+	Conditional	Yes
hand-held instruments			−	0		up to days	− to +	Conditional	Conditional
special instruments			+	+	+	up to weeks	+	Yes	Yes
flicker meter	?	+				up to weeks	+	Yes	/

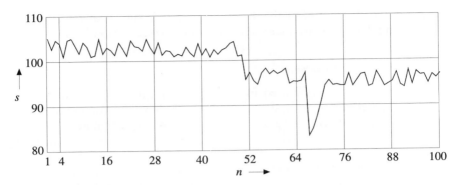

Figure 5.9   Sequence of measured values s[n]

Because of the problem that measurements of system perturbations and voltage quality must usually extend over a long time period in order to enable an assessment or analysis, the usefulness of individual measurement results is extremely limited. This means that a great number of individual results have to be reviewed and assessed. For this reason, a series of simple methods which further process the actual measurement results for different purposes is used to evaluate and compress measurement results.

In particular, the following method for processing a time sequence $s(n)$ is considered very important. The time sequence can, in a way, represent each basic measured value of a measurement. At this point it is unimportant whether this is a voltage or a harmonic component.

If one considers the example time sequence $s(n)$ in Figure 5.9, the mean-value generation is a frequently-used method for smoothing the measured signal. It can be in the form of simple mean-value generation. In addition to this form, methods are also used whereby the root mean square value or geometric mean value are determined. In the latter case, for example, the long-term flicker disturbance value P_{lt} is determined.

The r.m.s. values of voltage and current are calculated using the root mean square value.

Arithmetical mean value:

$$\overline{S}_A = \frac{1}{N} \sum_{i=1}^{N} \qquad\qquad (5.2)$$

Root mean square value:

$$\overline{S}_Q = \sqrt{\frac{1}{N} \sum_{i=1}^{N}} \qquad\qquad (5.3)$$

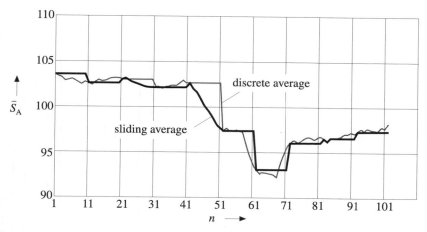

Figure 5.10   Sliding and non-sliding mean value calculation

Cubic mean value:

$$\overline{S}_K = \sqrt[3]{\frac{1}{N}\sum_{i=1}^{N}} \tag{5.4}$$

The deciding factor in the question of whether the data volume reduces during mean value generation or remains constant is whether the mean values are formed as 'sliding' or 'non-sliding' (see Figure 5.10). In the case of sliding mean-value generation the volume of data remains constant. This means that for each new value the first value of the interval under consideration is omitted and a new one added. If the mean value determination is non-sliding, the achieved reduction in the volume of data depends on the length of the mean-value generation interval. In the case of mean values which are determined non-sliding, the signal course obtained depends, in some circumstances, on the starting point of the mean-value generation. Mean-value generation smooths a signal.

A further step in describing a measured value sequence is the relative frequency. The relative frequency shows how many measured values, relative to the total number of measured values, of a sequence lie within a specific amplitude class. Figure 5.11 shows the relative frequency for the signal characteristic in accordance with Figure 5.9. With the relative frequency, the time relationship of the individual values of the output sequence are lost while sequences of any length can be concentrated in a limited space. The area in which the amplitude values and their distribution are located can be seen at a glance.

The following applies for the frequency:

$$h[n] = \sum_{i=0}^{n_{max}} \frac{s_i}{s_i(s_i = s_n)} \tag{5.5}$$

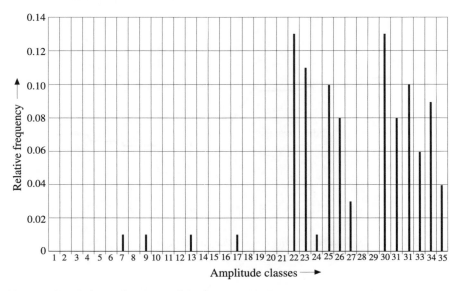

Figure 5.11　Relative frequency of the sequence s[n]

The relative frequency can also be calculated from the frequency $h(n)$, as follows:

$$h_{rel}[n] = \frac{h[n]}{\sum\limits_{i=0}^{n} h_i} \tag{5.6}$$

The measured values can also be described by determining the relative cumulative frequency. This is determined from the relative frequency by the summation of all the relative frequencies which are greater than, or equal to, the amplitude value under consideration. Figure 5.12 shows the characteristic of the relative cumulative frequency of the amplitude of the $s(n)$ time sequence.

The relative cumulative frequency can be determined purely formally in accordance with Equation (5.7).

$$C[n] = \sum\limits_{i=n}^{n_{max}} h_{rel}[i] \tag{5.7}$$

The form shown in Figure 5.13 is generally chosen for system perturbations.

From the course of the relative cumulative frequency, the amplitudes for different time durations can be read, relative to the overall measurement time period. Determination of the relative cumulative frequency is also, for example, used in the P_{st} disturbance evaluation method.

The 95% cumulative frequency value is used to check the compatibility level of harmonics. This means that relative to the interval being considered, the corresponding measured value of the voltage harmonic content must lie below the compatibility level for 95% of the examined interval. The measurement time

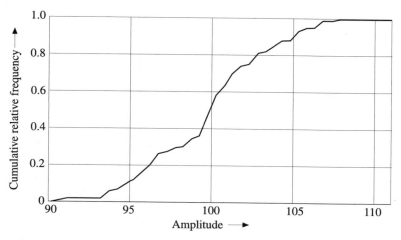

Figure 5.12   Cumulative relative frequency (normal representation)

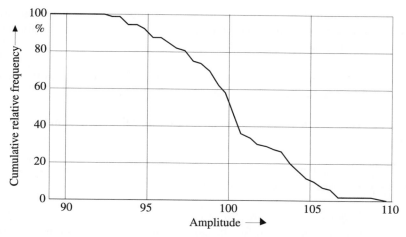

Figure 5.13   Cumulative relative frequency (representation for perturbations)

period is to be matched to the load cycle and therefore its duration cannot be predicted. Figure 5.14 is an example of the course of the cumulative frequency of a characteristic quantity (fundamental component). The 95% and the 99% cumulative frequency values are entered in this example.

5.3.2 Measuring and evaluation methods

To build up a wide base for successful measurements requires a wide variety of measuring methods, as well as suitable ways of evaluating and organising the various results of short- and long-term measurements.

A summary of the results of the direct physical measurements, combined with

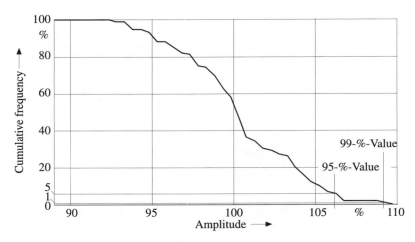

Figure 5.14   Example for cumulative frequency of a characteristic quantity

the sequential quantities derived from them, provides a variety of individual items of information with different content. These consist of the following quantities (see also sections 2.2.1 and 3.4).

Voltage and current in the form of the instantaneous value These quantities can be measured directly and are stored in time sequence form.

Voltage and current r.m.s. value These quantities are calculated from the particular instantaneous values.

Active, reactive and apparent power, power factor These are calculated either from the r.m.s. values or from the instantaneous values.

Harmonic components for current and voltage according to amount and phase Fourier transformation is used for the calculation. Sequential quantities can, in turn, be calculated on the basis of these quantities.

Angle of the harmonic components relative to the fundamental component

Angle between voltage and current of the harmonic components

Harmonic active power and harmonic reactive power

Total harmonic distortion factor (THD)

Weighted harmonic distortion for inductances

Weighted harmonic distortion for capacitances

Partial weighted harmonic distortion

Short-term flicker distortion value P_{st}, A_{st} Calculated from the voltage instantaneous values using the flicker calculation algorithm.

Long-term flicker value P_{lt}, A_{lt} Calculated from the short-term flicker distortion values.

Degree of unbalance of voltage and current Calculated from the fundamental components of voltage and current.

In addition to these quantities or their time sequences, it must be possible to determine the relative cumulative frequency values from the particular time sequences for all values which are being considered with regard to compatibility levels. In this case it is advantageous if the frequency limit can be freely stipulated. In any case, the values of the 95% cumulative frequency must be determined because the compatibility levels refer to this value. To properly determine the 95% cumulative frequency values relative to compatibility levels, the measurement time period must be chosen in advance so that it corresponds to a load cycle or a multiple of this time period. Often there is no information available in advance regarding the load cycle, so it is advantageous if the measurement segments used for the actual assessments can be freely stipulated.

5.4 Accuracy

5.4.1 Algorithms and evaluation

The accuracy of a measurement, considered over the complete measurement sector including evaluation, statistics and display, is subject to different error influences. The actual measuring accuracy in the calculation when acquiring current and voltage measured values is considered in section 5.4.2. The assessment of the voltage quality is, due to the frequency range in which the various effects occur, only possible by using several measurement functions. Each measurement function is itself subject to various limitations.

The harmonics analysis is, on the one hand, band-limited by the sampling frequency and, on the other hand, the window end from which the measured values are taken has an effect on the measurement result. If the harmonics level in a data block being analysed changes quickly, the resulting measurement will show large deviations. This particularly affects current measurements, such as current measurements on arc furnaces. To obtain accurate measurements the harmonic amplitude at a window of, for example, 160 ms or 200 ms must be almost constant for this time period. Disturbing influences occur when measuring flicker if very large voltage changes or voltage sags or voltage interruptions occur, because these cannot be depicted by the flicker meter algorithm.

It is desirable to minimise quantising influences to facilitate evaluation. This means that quantising in the magnitude of the measuring resolution is appropriate for processing measuring results in the statistics functions. Thus, at a measuring resolution of 0.1% of the nominal value, subdivision should also be made with categories of this magnitude.

5.4.2 Instrument and isolating transformers, current clamp

The desired or required measuring accuracy depends on the purpose of the measurements. The greatest measuring accuracy is required if the valid compatibility levels obtained from an assessment of the measuring results are used as reference values. In these cases, where consequences associated with costs for the power supply company or customer may be derived from the measurements, the measuring instruments used must comply with a defined accuracy range. Furthermore, in these cases the metrological boundary conditions and methods of analysis must also comply with the stipulated models, which guarantee equal treatment and reproducible measurement results. For this reason, the standards for the design of measuring instruments for the measurement and assessment of system perturbations specify the essential parameters which affect the measurement results. The block diagram of the measurement instrument to be considered is shown in Figure 5.15.

Particular attention must be paid to measuring accuracy where the measurements are taken via measuring transformers.

According to EN 61000–4–7 (VDE 0847 part 4–7), the transmission properties for voltage and current transformers of the various voltage levels should be assessed as follows.

Figure 5.16 shows three typical examples for the transmission behaviour of current clamps for use in a low voltage system. These are current clamps which convert the measured current into an equivalent voltage (A) and also pure current clamps (B). The transmission behaviour of a Rogowski measuring coil is also shown (C).

Low voltage: voltage and current transformers are generally well suited.
Medium voltage: voltage values with 5% measurement uncertainty at approx. 1 kHz; Angular error < 5° up to approx. 700 Hz

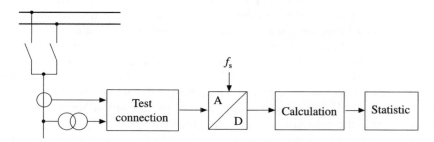

Figure 5.15   Principal structure of a measurement line

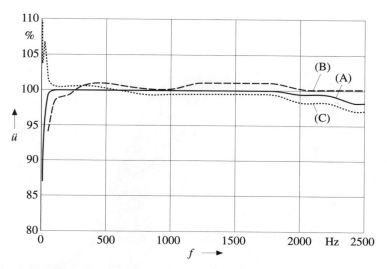

Figure 5.16a   Current clamp, curve of amplitude versus frequency

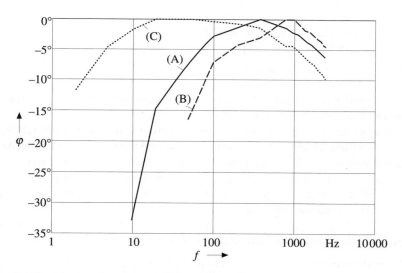

Figure 5.16b   Current clamp, curve of phase versus frequency

High voltage: voltage transformers which are very suitable up to approx. 500 Hz

Highest voltage: voltage transformers not suitable above 250 Hz

See Figure 5.17 for reference.

When measuring the instantaneous values which are assessed in the time range, the sampling frequency, amplitude resolution, linearity and bandwidth

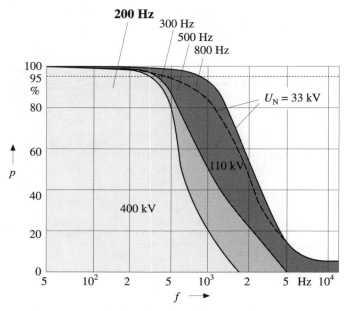

Figure 5.17   *Cumulative frequency of error for VTs*
Number of measured VTs: 41
——————— *measurement error 5%*
– – – – – – *measurement error 5°*

are the essential parameters. This also applies, in a similar way to harmonics measuring instruments (see Figure 5.18).

Of course, the type of windows for the measured values and the block length of measured data which must be used to determine the harmonics values have a decisive influence on the measurement results. The block length is particularly important when measuring harmonic levels which are not time constant, and not only affects the accuracy of the absolute value of the harmonics, but also has a quite considerable effect on the angular accuracy.

The following requirements apply to the performance of 'standard-compliant' measurements of harmonics EN 61000–4–7 (VDE 0847 part 4–7).

- The analysis interval must be matched to the equipment application.
- The measuring accuracy must be sufficiently high (class A for test bay measurements, class A or B for field measurements).
- The angular error must be less than ±5° or less than $h \times 1°$ (h is the harmonic order).

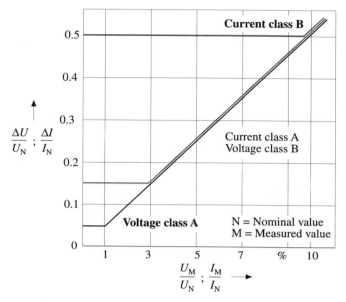

Figure 5.18   Accuracy requirements for harmonic measurements

5.5 Use and connection of measuring instruments

5.5.1 Low voltage system

In a low voltage system the connection of measuring instruments is usually straightforward. The voltages can be measured without the use of measuring transformers. The current can frequently be measured without difficulty using current clamps. The disturbing effect of unknown transmission functions is thus precluded. When current clamps are used it is also unnecessary to disconnect any current transformer circuits. Figure 5.19 shows the possible measuring and load situations in a low voltage system.

The measuring system is equally well suited for measuring harmonics and flicker. The phase-to-earth voltage and phase currents can be set in a direct relationship to each other. The phase-to-earth voltages are the measured variables suitable for flicker measurement because the lamp is also supplied via the phase-to-earth voltage.

5.5.2 Medium and high voltage systems

Measurements in medium and high voltage systems can only be made through measuring transformers. The fitting of special instrument transformers with known transmission functions is not possible in the majority of cases of all measurements. The possible measuring and load situations are shown in Figure 5.20.

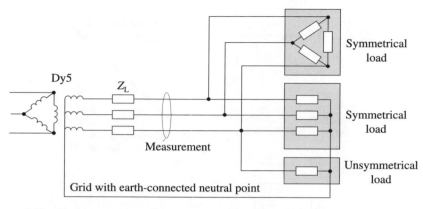

Figure 5.19   Connection of measurement system in a low voltage system (TN-System)

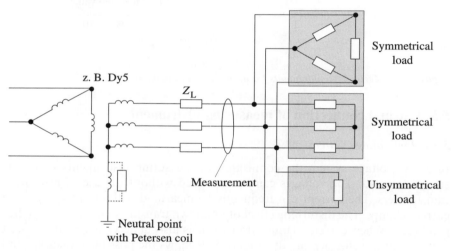

Figure 5.20   Connection of measurement system in a medium voltage system

This means that when assessing the measurement results the notes in section 5.4.2 regarding the effect of the transmission function of the transformer should be taken into account, particularly when assessing the measurements of harmonics. The transmission function does not play such a significant role in the assessment of the measurements of flicker.

In contrast to the connection of measuring instruments in low voltage systems, the connection of the measurement inputs in medium and high voltage systems can no longer be freely chosen. Depending on the construction of the transformer panels, the following various situations may arise in these cases (see Figure 5.21).

Fully-instrumented transformer panels with three voltage and three current transformers are relatively rare in medium voltage systems. For the measurement

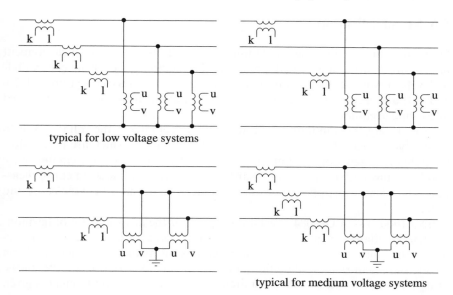

typical for low voltage systems

typical for medium voltage systems

Figure 5.21   C.T. and V.T. connections in low and medium voltage systems

of current, it is of lesser significance whether two or three current transformers are available. A missing current can, if necessary, be determined by 'computation' from two measured currents. This can also be achieved on the measuring instrument by a suitable connecting circuit.

The situation is rather more complex when measuring the voltage. It is desirable, when measuring harmonics, to measure the phase-to-earth voltage as well as the phase currents. In this way, information which can be assessed on the angular relationships between the harmonic voltages and the harmonic currents is also obtained. It is possible to convert the phase-to-phase voltages to phase-to-earth voltages via an artificial neutral point. However, this corresponds to the actual conditions only if the three-phase system does not have a zero sequence system. Furthermore, the artificial neutral point leads to a compensation of the harmonics of an order corresponding to a multiple of three.

The phase-to-phase voltages are of interest when measuring flicker in MV systems. Zero phase-sequence occurrences have no significance with regard to the supply of the loads in medium voltage systems. It is particularly important that the phase-to-phase voltages should be measured in medium voltage systems with earth-fault compensation because then any change in the zero sequence voltage shows up in the measurement result. Any change in the system can in this case lead to changes in the zero sequence system.

5.6 Standardisation

Standards stipulate binding regulations for various aspects of the measurement of system perturbations or voltage quality. The standards cover the compatibility levels for voltage fluctuations, unbalance and harmonics (see sections 2.4, 3.6 and 4.4) and also define the measuring and assessment methods and required measuring accuracy.

Increasingly, the standard specifications no longer refer to the voltage levels to be set, but instead define groups of emitted interference which are permissible for individual items of equipment or groups of equipment. This circumstance also has a favourable effect with regard to the metrological assessment of measurements because an appropriate differentiation can be made between cause and effect.

EN 61000–4–7 (VDE 0847 part 4–7) applies for the measurement of harmonics. This standard specifies the minimum requirements for instruments for measuring harmonics. It contains recommendations on methods of calculation, on the measurement range and on the statistical calculations. It also specifies measuring parameters and accuracy requirements. The accuracy requirements given in section 5.4.2 are taken from this standard.

Instruments for measuring flicker are described in EN 60868 (VDE 0846). This standard contains the possible measuring methods and algorithms for the measurement of flicker. Test sequences which stipulate the measuring accuracy are also given.

5.7 Characteristics of measuring instruments

Despite the features which a measuring instrument for the measurement of the properties of the voltage quality must have specified in standards, there is still some degree of tolerance left for the design of measuring instruments.

The properties which are specifically required depend to a large extent on the intended use of the measuring instruments. These objectives can be very varied, for example:

- assessment of voltage quality,
- compilation of a harmonics register,
- determination of basic values for calculations,
- performance of comparison measurements,
- determination of emitted interference,
- analysis of the causes of interference,
- checking and assessing of countermeasures,
- design and layout of equipment.

If some of the characteristics of measuring instruments are considered, it can be seen that their importance varies depending on the purpose of the measurements. A few of these characteristics are listed and explained in the following text.

Table 5.2 Measurement inputs

Voltage	100/√3 V	Measurement range factor 1.4
	100 V	Measurement range factor 1.4
	230 V	Measurement range factor 1.4
	400 V	Measurement range factor 1.4

Other measurement ranges via transformers or scalers

Current	1 A	Measurement range factor 2
	5 A	Measurement range factor 2

Other measurement ranges via intermediate transformers

Measurement inputs With regard to the measurement inputs, a suitable design of the measurement ranges is important, in addition to the number of channels of the voltage and current measurement inputs. For measurements in electrical power supply systems, the value ranges are distributed as shown in Table 5.2. Measuring instruments used in electrical power supply systems must have a sufficiently high overload resistance, so that they do not suffer damage in the event of a system fault.

Any number of measurement channels can be found on the various measuring instruments. Four voltage and four current measurement channels are certainly sufficient. This enables a three-phase system to be completely measured and it is still possible to include the zero sequence system in the measurements in the low - voltage system. Three measurement channels for voltage and current are still adequate, even though a longer measurement time period is required for additional investigations of the zero sequence system. Instruments with only one channel for current or voltage are significantly limited. When only the current or the voltage can be measured at one time this makes it distinctly difficult in many cases to analyse the relationship between cause and effect.

Measurement functions A consideration of desirable measurement functions leads to the conclusion that, in addition to the harmonics analyser and flicker meter as special measurement functions, an oscilloscope function performs useful services. Because special effects apply to only one part of a system period it could, in fact, be shown with the first-named measurement functions that a precise measurement or analysis is not possible for machine start-ups or commutation processes. If these considerations are extended to features which occur for durations of between only a few periods and a few seconds, the standard transient recorder is in many cases the only instrument for the more precise analysis of certain phenomena.

Bandwidth The bandwidth of the measuring instruments depends on the purpose. According to the stated standard (EN 61000–4–7; EN 60868), it can be determined that harmonics analysers must have a minimum bandwidth of

2.0 kHz. These instruments can then measure harmonics up to the 40th order. A bandwidth of 1.25 kHz is sufficient for many investigations. Nevertheless, harmonics of the 25th order can still be detected. For many measurements which extend into the area of disturbance analysis and fault diagnosis it is best to have the largest possible bandwidth. Instruments with a 2.5 kHz or 3.0 kHz bandwidth, which can then measure the harmonic components up to the 50th or 60th order, frequently provide interesting additional information. A flicker meter does not require such a large bandwidth. In this case values of 0.4 kHz to 1.2 kHz are sufficient, depending on the sampling frequency.

If, however, the instruments are provided with the functionality of an oscilloscope or transient recorder, the bandwidth must be as large as possible. If, for instance, commutation oscillation with an oscillation frequency of 4 kHz is examined, this signal characteristic can be detected even at a sampling frequency of 8 kHz, but for a clear depiction a sampling frequency of approximately 40 kHz is necessary (ten times the signal frequency).

Measurement time period For measurements of voltage quality and system perturbations in electrical systems of the public supply system it can be assumed that a complete measurement period with a duration of one week will be used. Measurements in industrial systems or measurements which are specified by a prepared measurement schedule, whereby, for example, specific system states can also be appropriately set, require considerably shorter measurement periods.

Data recording Data recording for long-term measurements must in most cases be performed with an averaging interval of 10 minutes, in order to ensure standard compliance.

If the dynamics of the recorded measured quantities are of interest, a shorter averaging interval is useful. At an interval of one minute, 1440 measured values are obtained over a 24 hour period. This means that a measurement covering one week consists of 11520 measuring intervals. The measuring instruments and the corresponding evaluation program must have sufficient capacity for these quantities. A very short measuring interval is important for special investigations. This should just be seconds.

5.8 Performance of measurements

Measurements are generally taken for quite different reasons. The first step in the preparation of a measurement is to define the objective of the measurement. The question 'What is to be achieved?' must be answered. The next step is choosing the measurement site and specifying the measuring instruments. When the measurement site and measuring instruments have been established, aspects of the connection of the measuring instruments must be considered. Particularly where measurements are to be taken over a long time period, the measuring instruments should be located in order to cause the least disturbance. In switchyards, particularly in customer systems, this may not be quite so simple. The installed measuring set up must also be adequately protected against unauthorised access.

5.9 References

1   UCTE: Statistical Yearbook
2   KÖHLE, S.: 'Ein Beitrag zur statischen Bewertung von Flickern (A contribution to the static assessment of flicker)' *Elektrowärme International*, 1985, vol. 10, pp. 230–239
3   MOMBAUER, W.: 'Neuer digitaler Flickeralgorithmus (New digital flicker algorithm)' etz archive 10, 1998, pp. 289–294

Chapter 6

Countermeasures

6.1 Assignment of countermeasures

Countermeasures to compensate for system perturbations can be applied at different points in the electrical system (see Figure 6.1).

The various measures are dealt with in the following text, with reference to Figure 6.1.

6.2 Reduction of the emitted interference from consumers

When dealing with compensation of harmonics in a consumer network connected through a transformer to the supply system, the transformer circuits shown in Figure 6.2 are particularly suitable for compensation for the third order harmonic. Because the third order harmonic forms a pure zero sequence system with regard to its representation in symmetrical components (see section 1.4.3), the use of transformers which permit no transmission of the zero sequence system from the primary to the secondary side enables this harmonic to be compensated for. Because of the asymmetries of the transformer construction, the impedance of the zero sequence system is finite. This causes the transmission of the third harmonic from approximately 10% to 15% via the transformer windings.

If the supply system is connected at the consumer end through converters, e.g. for drives, the control process can be optimised to achieve a low harmonic distortion.

This can be achieved by two six-pulse circuits. By means of a three-winding transformer the current components are superimposed in such a way that the resulting supply current in particular contains a component of the twelfth harmonic with a correspondingly-low amplitude (see Figure 6.3).

Line-commutated converter circuits can operate only in block mode because the thyristors require commutating voltage (Figure 6.4a). Self-commutated

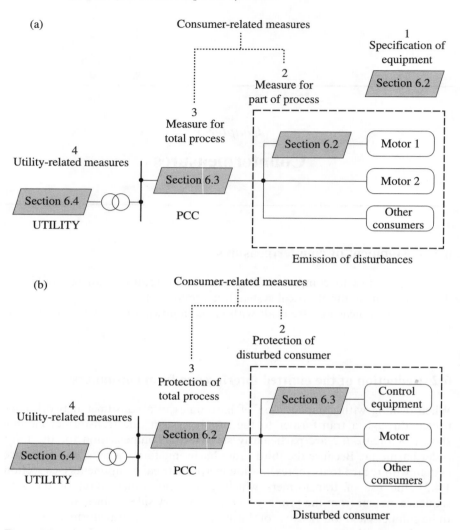

Figure 6.1   a)   Countermeasures related to disturbance emission (disturbance source)
b)   Countermeasures related to disturbed consumer (disturbance drain)

converter circuits on the other hand require no commutation voltages and are therefore capable of high-pulse operation (Figure 6.4b). This enables the harmonic distortion to be significantly reduced, as is shown clearly by the current spectra in Figure 6.4.

The effect of form factor, frequency and change in voltage amplitude on the flicker value from the point of view of a reduction of flicker on the process side has been shown in section 3.3.3.

A measure frequently used to limit voltage sags in drives is the integration of starting-current limiting (starter with autotransformer, starting via a resistor or

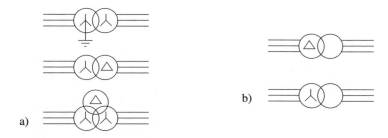

Figure 6.2   Switching arrangement of transformers
   a)saturation of transformer, avoidance of 3rd harmonic in secondary voltage
   b)non-linear load, avoidance of 3rd harmonic in primary current

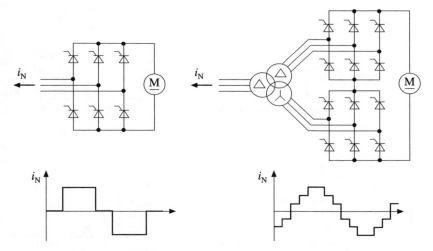

Figure 6.3   Arrangement of six- and twelve-pulse rectifier

reactor, part-winding starter, star-delta starting) or soft starters. In this case, however, the harmonics caused by these devices in the partial-load range must be considered [1].

Where arc furnaces and welding machines are used, the choice of d.c. furnaces or d.c. welding machines and the inhibiting of individual part-consumers can also lead to a reduction in the level of the voltage sags. On welding machines, the flicker form factor can be changed by choosing a waveform of the welding impulse to reduce the flicker. The same can be achieved for furnaces by changing the electrode control, whereby this measure can also be used to change the flicker frequency.

Any impairment of the production process and any expense attributable to installation or protection measures must be taken into account when assessing these measures.

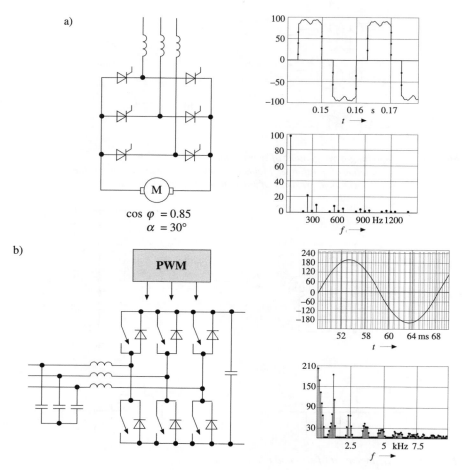

Figure 6.4   Control strategies for converter circuits

a) line-commutated strategy
b) self-commutated strategy

6.3 Consumer-related measures

6.3.1 Filter circuits

Sections 6.3.1, 6.3.2, and 6.3.3 are based mainly on the papers referenced under [7] and [8]. Capacitors without blocking reactors form a parallel resonant circuit with the supply system inductivities (see section 2.3.3). By resonance amplification they can thus contribute to an increase in the harmonic distortion of the power supply system. The resonance amplification acts both on the disturbance level in the transmission network and also on the harmonic currents generated in the customer network. Excessive harmonic disturbance levels not only cause

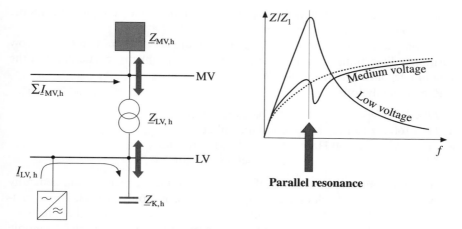

Figure 6.5   Power system diagram and impedance curve for parallel resonance circuit

capacitors to be very quickly overloaded due to their frequency characteristic, but can also disturb the operation of other consumers. A higher design voltage for capacitors is therefore not a solution. For a resonance amplification to be critical, a resonance close to a typical harmonic frequency is sufficient (see Figure 6.5).

In addition to parallel resonance, where capacitors without blocking reactors are installed, series resonance (see section 2.3.3) to the primary power supply system can also occur. In the case of series resonance (Figure 6.6) the current is essentially drained, i.e. no current amplification occurs. Of course, at this resonance the voltage harmonic content at the installation level of the capacitors is also magnified and operating conditions can thus result which are completely unacceptable for the customer if resonance close to a typical harmonic

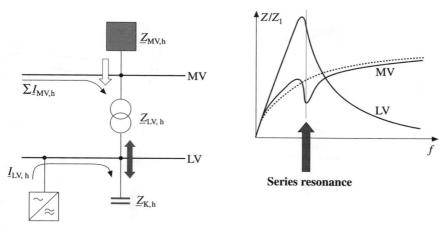

Figure 6.6   Power system diagram and impedance curve for series resonance circuit

frequency occurs. Supply system resonances in the area of typical harmonic frequencies often trigger circuit breakers.

In compensation systems blocking devices are used, even at relatively low harmonic distortion, in order to avoid resonance problems, particularly in the area of telecontrol frequencies (*TF*).

Recommendations

- *TF* < 160 Hz: 7% blocking reactor
- *TF* = 160 Hz to 190 Hz: combined filter
- *TF* = 190 Hz to 250 Hz: 12.5% blocking reactor
- *TF* > 250 Hz: 7% blocking reactor

Critical resonances in the area of typical harmonic frequencies can generally be avoided by suitable blocking of capacitors. This not only protects the capacitors from overload but also protects the complete system from the effects of resonance amplification. Blocking reactors prevent the increase in the harmonic distortion due to resonance amplification while at the same time improving the quality of the supply system because blocking the capacitors has a drain effect on harmonic currents. When choosing the type of blocking, the telecontrol frequency used in the network area of the relevant power supply company must also be taken into account, to avoid impermissible influencing of the telecontrol.

In filter circuit systems (Figure 6.7), the individual filter circuits are each tuned to a typical harmonic frequency. In this case the property of the series resonance is appropriately used to amplify the drain of the harmonic currents and thus substantially improve the network quality. Such filter circuit systems are actually considered to 'clean' the customer power. When designing such systems, the harmonic distortion of the primary network must, however, not be forgotten, because the harmonic currents coming from there simply cannot be ignored.

In the case of the fifth harmonic, a quite considerable additional loading must be expected because of the wide use of consumer electronics (TV).

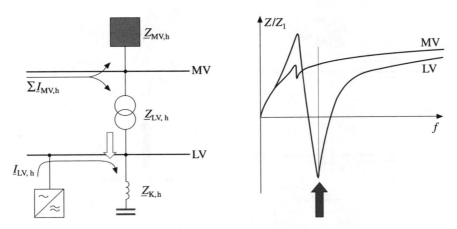

Figure 6.7   Power system diagram and impedance curve for filter circuit

Where filter circuit systems are connected directly to the supply network, the filter circuits act on all harmonic sources in the customer network. Furthermore, a high harmonic distortion introduced from outside results, must be allowed for when designing the system. Decoupling with current limiting or commutating reactors not only substantially increases the drain effect on the primary harmonic generators but also distinctly reduces the harmonic distortion from outside. A network decoupling is thus particularly useful if there is only a low compensation requirement or if several filter circuit systems in the same network cannot be coupled because of differences in design (Figure 6.8).

Filter circuits with the same resonant frequency, which are designed for operation in parallel on the network, must be coupled by switchable equalising conductors, in order to avoid different harmonic distortions and thus an over-load of the filter circuit (Figure 6.9). Without such a coupling, tolerances in the tuning would lead to very different harmonic distortion, whereas equalising conductors result in an equal relative harmonic distortion. Another con-sequence is that filter circuits of different design cannot be coupled. In this case adequate decoupling must also be provided, if necessary.

Filter circuit systems which are largely decoupled from the network by commu-tating reactors require no additional compensating lines when operated in paral-lel in the same network. The decoupling is itself sufficient isolation. Such filter circuit systems can be considered exclusively as assigned to the corresponding harmonic generator, the harmonic distortion from outside is comparatively low compared to systems connected directly to the network (Figure 6.10). A filter circuit system can, of course, be combined with a direct network connection (e.g. installed for several harmonic generators) to systems decoupled from the network (which are designed in each case for only one harmonic source) without coupling by equalising conductors being necessary.

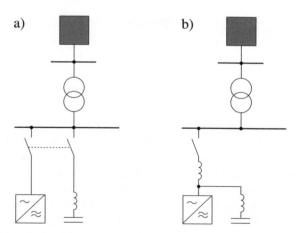

Figure 6.8   Connection of filter circuits
a)   direct coupling to power system
b)   indirect coupling to power system

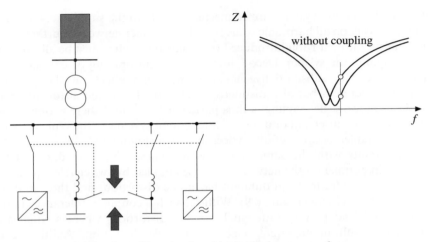

Figure 6.9   Coupling of two filter circuits with identical resonance frequency

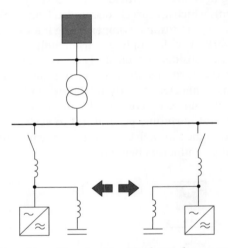

Figure 6.10   Decoupling of two filter circuits with identical resonance frequency

Multistage filter circuit systems must always be constructed in a gapless sequence for the typical harmonic frequencies. It is also not possible to dispense with filter circuits for this frequency for 12-pulse converters because of the ever-present disturbance level of the fifth and seventh harmonic. Multistage filter circuit systems are always switched in with increasing sequence and switched off in the opposite sequence, to avoid critical resonance amplifications and the resulting system overloads. An incorrect switching sequence must therefore be prevented by interlocking that is also effective even if a filter circuit fails.

There are various protective devices for filter circuit systems (Figure 6.11), such as:

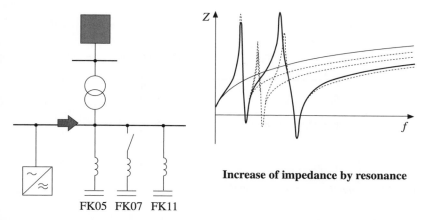

Increase of impedance by resonance

FK05   FK07   FK11

Figure 6.11   Switching sequence of three filter circuits with different resonance frequencies

- interlocking of the switching sequence,
- temperature control of filter circuit reactors,
- safety monitoring,
- measurement of harmonics distortion,
- detection of asymmetry of the filter circuits.

For all monitoring systems, there is a facility to indicate the status, provide a warning and shut down the system.

6.3.2 Dynamic reactive power compensation

Flicker disturbances or voltage fluctuations are caused mainly by changes in the reactive power, but changes in the active power can also cause the supply voltage to change. The disturbing effect of both the reactive power flicker and the active power flicker can be rectified by compensation. The most important factor in this case is a system reaction time which is as short as possible. In contrast to conventional control systems the compensation demand is preferably determined by measuring the load current. This open-loop method substantially reduces the control lag. In special cases a direct control is also installed to deal with flicker disturbances. Figure 6.12 shows the topology of such a dynamic reactive power compensation system.

6.3.3 Symmetrical connections

Network asymmetry can be rectified by reactive power compensation and symmetries corresponding to the Steinmetz circuit (see Figure 6.13). If several high-power, two-phase loads are operated independent of each other in a network, a considerable effort may be required to remove disturbing network asymmetries.

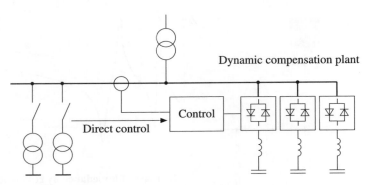

Figure 6.12   Dynamic reactive power compensation

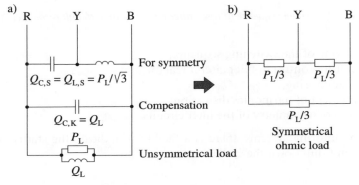

Figure 6.13   Circuit connection for symmetrical power system connection of
*   unsymmetrical load*
*   a) electrical diagram*
*   b) load seen from the power system*

6.3.4 Active filters

The filter circuits described in section 6.3.1 are also known as passive filters, designed to compensate for defined harmonics. The problems associated with these are also dealt with in the aforementioned section. A completely different mode of operation is created by so-called active filters, with self-commutated converters being used to compensate for distortive reactive power by injecting the negative harmonic spectrum. Seen as a presentation of a model, the passive filter drains an nth current harmonic due to resonance. The active filter, on the other hand, causes an addition of the negative harmonic currents. These functions are compared in Figure 6.14. An important feature in this case is allowing for the switching frequency as a limiting factor.

The suitability for use of active filters to compensate for harmonics is examined first. In this case the network connection used with self-commutated converter circuits is particularly significant. The use of active filters to compensate for

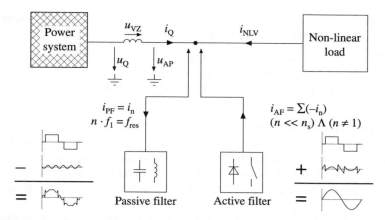

Figure 6.14   Connection of shunt-connected active filter and passive filter

low-frequency system perturbations, and particularly flicker compensation, are then dealt with in conjunction with a consideration of energy storage devices.

The active filter described is normally designed to function as a parallel filter as shown in Figure 6.14 (direct compensation of current harmonics) with it being possible to use the current or voltage at the connecting point as a control variable. The active filter can also be designed to function as a series filter, as shown in Figure 6.15 (isolation of voltage harmonics). In this case the active filter is used as a voltage source to screen sensitive consumers from system perturbations. It can also be used to compensate for voltage reduction and voltage sags. However, an energy storage device e.g. battery is then required.

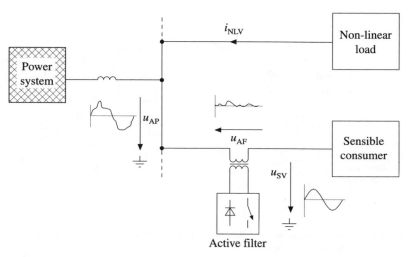

Figure 6.15   Connection of series-connected active filter

In conjunction with the construction of active filters, the following can, in principle, be used as self-commutated, directional valves.

- GTO   Gate turn-off thyristor,
- IGBT   Insulated gate bipolar transistor,
- Power MOSFET   Metal-oxide semiconductor field effect transistor,
- MCT   Metal-oxide semiconductor controlled thyristor.

The system converter circuit can be designed either as an *I*- or a *U*-converter. In the first case, a current source on the d.c. voltage end is used, and in the second case a voltage source. Industrial development of converter technology in recent years has clearly shown that the *U*-power technology has come to the fore for self-commutated circuits and thus is particularly suitable for use as an active filter.

Associated with this is the idea of a so-called Unified Power Conditioning System (UPCS) [2]. The following are the performance features of this pulse-width modulated IGBT converter:

- reduction in commutation notches,
- reduction in harmonics,
- reduction in voltage fluctuations (flicker),
- provision of reactive power.

This results in the topology for the UPCS shown in Figure 6.16. Depending on the required improvement in voltage quality, the UPCS power for operating systems with a multiple of the UPCS power is sufficient because the deviations of the voltage from its ideal value is only in the percentage range. A fraction of the total customer load is also sufficient for the reactive power provision.

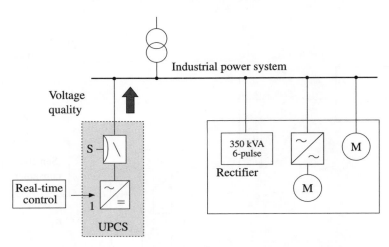

Figure 6.16   Principal diagram of Unified Power Conditioning System (UPCS)
         S, electronic switch
         UPCS converter in IGBT-technique

The control algorithm of the UPCS operates in the time domain, i.e. in real time. This means that each instantaneous voltage change can be corrected in fractions of a millisecond, whereas a control algorithm operating in the frequency range requires a few milliseconds to first analyse the disturbing phenomenon before providing control instructions for its correction. Such systems therefore have a lower performance in compensating for high-frequency system perturbations, compared with systems which provide control in the time domain. The design and use of a UPCS are described in detail in section 6.6.1.

For an active filter to be used for system perturbation compensation, it also needs, in addition to being connected to the system by self-commutated converter elements, an energy storage device for the provision of compensation active- or reactive power.

A capacitor is sufficient as an energy storage device for harmonics compensation. If flicker is considered as well, capacitors are generally no longer sufficient. For this, energy storage devices with a larger energy capacity, as described in the following, are required. In this case, the use of such systems is substantially more cost effective if other functions (stand-by power supplies, load management, stabilisation) can still be provided at the same time.

6.3.4.1 High-performance batteries

Lead-acid batteries presently dominate the field of stationary battery applications for electrical power supply. Their design varies depending on the application. Load management requires systems designed for a long life cycle, while particularly reliable systems are needed for the stand-by power supply area where discharge and subsequent charging operations rarely occur. Electrolyte circulation aids are used to prevent stratification of the battery acid during operation.

Nickel-cadmium batteries are particularly suitable where shorter discharge times and longer service lives are required but such systems may cause problems, particularly with regard to environmental compatibility and cost. Compared with lead-acid batteries, the increased reliability of the individual cells must be set against the greater number of cells required.

Sodium-sulphur batteries have been developed mainly for mobile applications and, compared with lead-acid batteries, have the advantage of a lower weight and smaller external dimensions. The chemical process within the battery takes place without the output of heat loss and the theoretical charging efficiency is thus 100%. The process temperature is 340 °C. Because the internal temperature rise is unacceptably high for large discharge currents, there is, at present, an extremely low power density for short-term storage. Because of the very high cost of sodium-sulphur batteries, these systems will not be significant in the immediate future for stationary use.

Overall it can be seen that lead-acid batteries will also continue in future to dominate as stationary batteries for power supply, particularly due to the demonstrable cost-effectiveness. Figure 6.17 shows the principle of operation of electrochemical energy storage for the lead-acid battery.

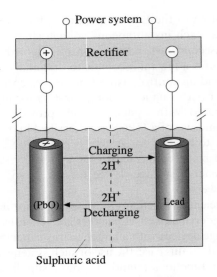

Figure 6.17   Electro-chemical process in a lead-acid battery

The reaction equations for the positive electrode are as follows:

$$PbO_2 + H_2SO_4 + 2\,e^- \rightarrow PbSO_4 + 2\,OH^- \tag{6.1}$$

and for the negative electrode are:

$$Pb + H_2SO_4 \rightarrow PbSO_4 + 2\,e^- + 2\,H^+ \tag{6.2}$$

Because of their specific discharge characteristics or internal resistances, the exclusive use of batteries to compensate for voltage fluctuations is not appropriate. But when used together with a suitable system connection, possibilities for multifunctional use of such overall systems in different time ranges result.

6.3.4.2 Superconductive magnetic energy storage

Because an extensive parallel circuit of capacitors is not without problems and a battery due to its dimensions and internal resistance is frequently not the optimum solution for a fast power pulse, a superconductive magnetic energy storage device (SMES) is necessary where high power is required for short periods.

In a superconductive magnetic energy storage device the energy is stored in the magnetic field of a superconductive coil. This requires temperatures of 4 K (metal superconductor) or up to 77 K (ceramic superconductor) to guarantee the superconductive state ($R = 0$) and thus the actual storage property of the system. Liquid helium (4 K), gaseous helium or liquid nitrogen (4 K to 77 K) is used for cooling. The superconductance also depends on the external magnetic field and the actual current in the conductor. There are also critical values here which may not be exceeded, otherwise the superconductive state is lost (so-called quenching, see Figure 6.18).

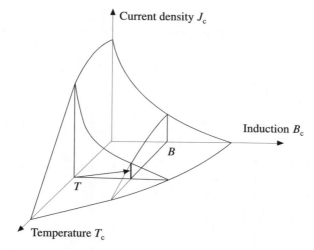

Figure 6.18   Operating diagram of a super-conducting circuit

Although ceramic superconductors have interesting values with regard to critical temperature, only very low critical currents can be achieved in these temperature ranges and thus, for technical and economic reasons, only metal superconductors can at present be considered. The basic design of such an SMES is shown in Figure 6.19.

The energy which can be stored in a coil as shown in Figure 6.19 is as follows:

$$E = \frac{1}{2} \int\int\int_{v} \vec{B}\vec{H} dv = \frac{1}{2} LI^2 \qquad (6.3)$$

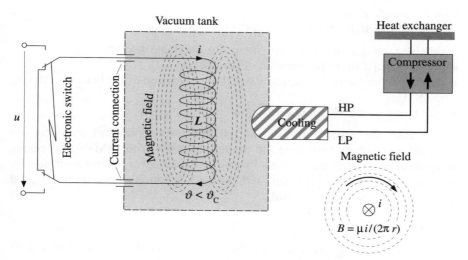

Figure 6.19   Super-conducting Magnetic Energy Storage (SMES)

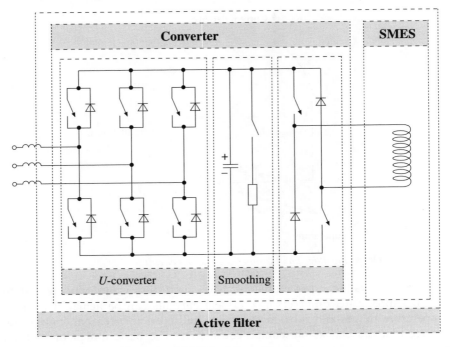

Figure 6.20   Electrical diagram of SMES with rectifier

In order for an SMES to also be able to utilise the advantages of a *U*-converter topology (see above), a link, as shown in Figure 6.20, is required between the impressed SMES and the line converter requiring a voltage of maximum constancy. Where used in a four-wire system this voltage, ideally, should be provided with a loadable neutral point.

An ideal link element should meet the following requirements:

- reliable provision of a constant voltage for the line converter,
- guarantee of a smooth operation of the SMES,
- decoupling of the high-frequency switching operation of the line converter by the SMES,
- protection of the SMES in the event of malfunctions.

In this case, the circuit arrangement and operation of the line converter are designed particularly to reduce the thermal loading of the SMES and ensure integration of protective functions. In addition, a loadable neutral point for use in a four-wire system is provided, without the line converter having to be fitted with a suitable secondary control.

6.3.4.3 Gyrating mass flywheel

In a gyrating mass flywheel (GMF), the energy is stored in the form of kinetic energy in a rotating flywheel.

$$E = \frac{1}{2}\theta\omega^2 \tag{6.4}$$

In operation, the flywheel is driven at variable frequency by an electric motor/ generator controlled by an inverter control. This converts the electrical energy to kinetic (rotation) energy which is stored in the flywheel and converted back into electrical energy as required.

To store a large amount of energy requires either a high speed angular velocity ω or a corresponding high moment of inertia θ to be applied. This energy is converted to electric current by a motor-generator unit and supplied to the consumer. At present, a distinction is made between low-speed storage devices running at approximately 3000 r.p.m. and high-speed storage devices with speeds of up to 20000 r.p.m. The high-speed systems are relatively more compact because of the lower requirement for the moment of inertia. However, the development of the high-speed systems presents a challenge because of the speed-related friction losses. The introduction of magnetic bearings has enabled the performance of gyrating mass flywheels to be substantially increased. Because in a magnetic bearing (superconductive magnets are also used) there is no contact between the moving parts, a large part of the operating and development problems associated with conventional bearings (ball and roller bearings, plain bearings and gas bearings) is avoided. There is still more development work to be done on high-speed systems and at present the only economically attractive systems available on the market are the low-speed systems.

The advantages of modern gyrating mass flywheels compared with conventional accumulators are a service life which is longer by a factor of at least 10^3, lower losses during long-term storage, higher output power during short-term storage and very good environmental compatibility.

The gyrating mass flywheel is thus, in principle, equally suitable for long-term storage and for short-term storage (fast power storage). This opens up many promising applications which cannot be economically met by the SMES or the high-performance battery. The topology of a system which is suitable for stand-by power supply and system perturbation compensation is shown in Figure 6.21 [3].

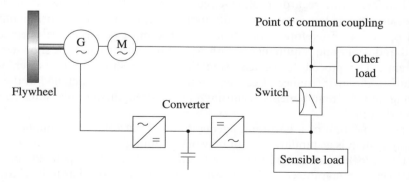

Figure 6.21   Gyrating mass flywheel

6.3.4.4 Comparison of various energy storage devices

Table 6.1 shows a comparison of the aforementioned energy storage devices for the application of stationary storage devices under consideration with regard to the compensation of system perturbation.

It can therefore be seen that at present the lead-acid battery, low-temperature SMES and low-speed gyrating mass flywheel in particular are economically significant. Batteries are particularly used for stand-by power supply applications, while gyrating mass flywheels and also SMES are worth considering as short-term storage devices to compensate for low-frequency system perturbations.

6.4 Measures related to power systems

6.4.1 Measures during network planning: system strengthening measures

The short-circuit currents, or the short-circuit power, can be considered to have both unfavourable and favourable properties. On the one hand fault currents can, under certain circumstances, cause high thermal or mechanical stresses in the power supply equipment, thus necessitating limiting these currents in conjunction with a mode of operation of the power supply equipment which affords the longest possible service life.

On the other hand, high short-circuit currents are desirable and, in many cases, even necessary in order to provide a safe activation criterion for the power supply protection and to guarantee selectivity. Furthermore, a small power system internal resistance (a high short-circuit power), is necessary if consumers with a heavily-fluctuating power requirement are to be connected, to ensure that system perturbations can be held within permissible limits. It must also be considered that the penalty for achieving a low power system internal resistance, including system strengthening measures, can be that suitable equipment has to be installed and that this is often associated with high costs and also with difficulties in obtaining legal approval.

Therefore, with regard to short-circuit currents, the principle 'as high as necessary, but as low as possible' still applies today both for network planning and operation. Therefore, a reasonable balance must be struck between technical and economic requirements with regard to the limitation of short-circuit currents. The procedure for assessing the existing short-circuit power, with regard to the maintenance of certain compatibility levels, is detailed in the following and the requirements regarding system strengthening measures can be derived from this.

The connection of a harmonics-generating new customer to a public power supply system is a standard task of network planning. The question of whether electromagnetic compatibility for all equipment to be operated on the system can be guaranteed under present or future conditions has to be clarified. Normally, changes in the status quo due to the connection of a new customer

Table 6.1 Characteristics of various energy storage devices [4]

	Lead-acid battery	Nickel-cadmium battery	Sodium-sulphur battery	LTS SMES	LTS SMES	GMF, high-speed	GMF, low-speed
Market readiness	High	Low	Low	Medium	Low	Medium	High
Typical starting time	2 min	2 min	30 min	15 s	15 s	15 s	15 s
Efficiency	85%	75%	98%	98%	98%	90%	90%
Losses due to	Self-discharging	Self-discharging	Reaction heat	Heat input	Heat input	Friction	Friction
Risks	Acid	Alkalis	Heat	Quenching field	Quenching field	Mass	Bearings
Anticipated cycles	1,500	1,000	1,000	100,000	100,000	1,000,000	1,000,000
Price in £ for 1 MW	33,333	133,333	—	666,666	1,000,000	100,000	33,333

are predicted by simulation calculations in order that the necessary measures can be implemented at the planning stage.

Of central importance is the superimposition of the harmonics caused by the new connection on the random basic levels already present in the system, which can be caused by low-power non-linear consumers.

The two following basic procedures are used.

- The arithmetical superimposition of the instantaneous values of the harmonic under consideration relative to the indirectly-coupled load—the result of such an assessment is on the safe side and, because of the lower cost for the input data, is easy to deal with.
- Taking account of the statistical nature of all harmonic currents and impedances with regard to amount and phase. A simulation process of this kind describes in detail by using distribution functions, the real relationships of the random harmonic voltage at all network nodes, but does not have any inherent safety margins. This method is more demanding with regard to the input parameters required.

Because of the random character of distributed small users, statistical methods are used for their simulation. Overall it is accepted that mathematical calculation procedures with sufficient accuracy for harmonics in public supply systems are available. These procedures are much more difficult to use because the cost of procurement of the necessary input data is sometimes high and also by their random character. The distributed small users present the greatest problem in this respect. The harmonic conditions in the public supply system are always determined by the load- and frequency-dependent impedances and the original harmonic currents of the small consumers. Because of the nature of the consumer, both are random variables and therefore accessible only by statistical means. The resulting harmonic conditions in the complete system at any amplitude and phase angle of the converter harmonics can be examined and violations of the compatibility level can be predicted.

Approaches of this kind to allow for certain consumers who generate harmonics can also be included in long-term planning. This means that on the basis of the actual situation and the anticipated development with regard to harmonics generators it can be determined whether the permissible compatibility level in the system is being maintained, the time point at which any problems can also occur in this connection and what measures with regard to system (expansion) planning or other methods of compensation are to be taken, and when.

6.4.2 Measures with regard to system operation: short-circuit current limitation

To increase the short-circuit power in undisturbed operation without at the same time increasing the actual short-circuit currents which occur in the event of a malfunction, the use of so-called short-circuit limiters at various points in the electrical supply system is recommended. Particularly with regard to current discussions and developments, in conjunction with the liberalisation of power supply, it will be necessary to have a system whose use offers an optimum economic solution with regard to network planning and operation. There are many interesting possible applications for such equipment.

Voltage quality, system perturbations The supply to customers requires that for load flow reasons a system connection must be provided which has adequate transmission power from lines and transformers. However, the system perturbations (harmonics and flicker) of customer equipment are increasingly of influence in the design of the connection, as the aforementioned large power electronic equipment as well as large arc furnaces, welding machines etc. require connections with a high 'voltage rigidity', i.e. a high short-circuit power. In the past this has led, in individual cases, to customers having to be connected to a higher voltage level than was necessary for their energy requirement.

By coupling busbars via short-circuit current limiters, the short-circuit power in normal system operation could, in some cases, be approximately doubled without having to change the operating equipment. This meant that in these cases a connection to the primary supply system at a higher connected power level would only be necessary if such a supply would also be useful for load flow reasons.

Use of transformers with lower short-circuit voltage u_k Transformers (110 kV/ medium voltage) with a lower u_k can be used if short-circuit limiters are fitted in the outgoing transformer circuit. From the point of view of network perturbations (see above), this enables the short-circuit power occurring in the 10 kV systems to be again increased in individual cases by approximately 150% to 200%.

Overall, this presents the possibilities, shown in Figure 6.22, of fitting short-circuit limiters in the electrical supply system.

A common method of achieving sufficiently high attenuation of the short-circuit power in the event of faults in the medium voltage system is the installation of short-circuit reactors. The reactor shown in Figure 6.23a connects both sub busbars and in normal operation is almost de-energised. It does not markedly increase the system internal impedance until a short-circuit has occurred.

However, the use of reactors as shown in Figure 6.23b and 6.23c has a detrimental effect in that, although it increases the internal impedance in the event of a short-circuit, load current flows permanently through it in undisturbed operation. This means that their reactances are completely effective in normal operation and are thus detrimental to voltage stability. A further disadvantage is that the use of short-circuit reactors is limited to the medium voltage level.

The most effective measure for limiting short-circuit current at present is considered to be the so-called pyrotechnic short-circuit current limiters (see Figure 6.24). If a short-circuit occurs, these cause a contact to be blown open by an explosive capsule as soon as there is a rise in the current.

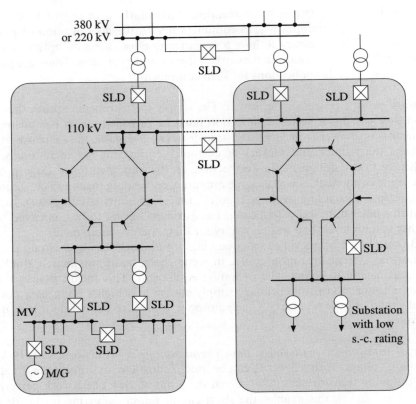

Figure 6.22   Application of short-circuit limiting devices (SLD) in power systems

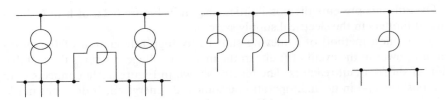

Figure 6.23   Application of short-circuit limiting reactors

The current is thereby commutated on a high voltage fusible link arranged in parallel to this contact, which then limits the current and quenches the arc. Because frequently only short-circuit limiting but no unselective shutdown is required, the short-circuit current limiters in such cases are arranged parallel to short-circuit reactors. In fault-free operation they are short-circuited by the short-circuit current limiters. The great disadvantage of these limiters is that their operating principle is not inherently safe, i.e. the current must be measured and evaluated, a trigger signal transmitted and an explosive capsule fired.

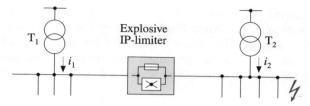

Figure 6.24   Application of an explosive short-circuit limiting device

The use of an optimised power electronic short-circuit current limiter using thyristor technology (Figure 6.25), which as a three-phase voltage controller can guarantee both starting-current limiting and also soft starting, enables the short-circuit current to be shut down within the first 1 to 2 ms after occurrence of the fault by an additional quenching circuit, thus ensuring immediate reavailability after rectification of the fault. In addition, the occurrence of overvoltages is avoided by a soft limiting of the short-circuit current. In a steady-state condition no system perturbations of any kind associated with the operation of the equipment occur [5].

Compared with other concepts of current limiting which require external triggers, a superconductive short-circuit current limiter (SSL) is able to operate with inherent safety due to the properties of the material of the superconductor, because the superconductor quenches if a current limit is overshot (changes from a superconductive to a normal conductive state). Another aspect of central significance is the regeneration process of the superconductor, i.e. the time taken by recooling after a quench until the superconductive state is again restored. The SSL in Figure 6.26 in this case represents a self-generating component which can

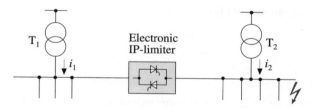

Figure 6.25   Application of an electronic short-circuit limiting device

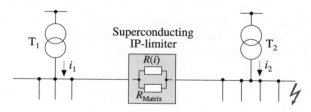

Figure 6.26   Application of a super-conducting short-circuit limiting device

return to its normal operating state. This is a major advantage of the SSL compared to short-circuit current limiting concepts where the subcomponents are irreversibly damaged after triggering the limiter and have to be manually replaced. It should of course be mentioned that, because of the technology, the SSL will not be of great economic interest in the short term.

6.5 Cost analysis

The main function of the cost-benefit model is to determine the relative potential of individual countermeasures to reduce system perturbations. The aim is to obtain useable, realistic results within a short time. There are various techniques for calculating the cost–benefit values, such as repayment or internal repayment rate (IRR). The cost–benefit calculation is performed using financial techniques which are generally accepted in the USA and Europe.

In the following, the generally-accepted cash value method according to Equation (6.5) is to be used as a basis for the cost–benefit calculation.

$$k_0 = \sum_{t=1}^{T} \frac{V_t}{(1+d)^t} \qquad (6.5)$$

where

k_0 is the cash value
T is the time period under consideration
t is the annual step
V_t is the value in year t
d is the annual interest

The benefit-cost rate (BCR) is thus

$$BCR = k_0/\text{IC} \qquad (6.6)$$

where

IC is the initial capital cost

Each of the parameters in the cash value calculation can have a significant effect on the final value of the benefit–cost rate. For this reason the values for T, d and V_t must be evaluated with care to ensure the values used are realistic. The typical annual repayment rate is set between 9% and 11%. This amount includes an inflation rate. Accordingly, an annual interest of 6% to 7% should be used. The benefit of a measure V_t is to be considered and determined specifically for each customer. The general investment parameters IC for system costs are subdivided into the following areas:

- installation costs,
- operating costs,
- maintenance costs.

There are frequently boundary zones to which certain costs should be assigned. However, only the final result is actually important. Therefore, how the costs are planned is not particularly significant provided all the costs are allowed for (and there is no double calculation). The following paragraphs provide information on how the detailed costs are shown.

System costs This item covers all costs which are directly connected to the hardware of the countermeasures, such as batteries, installation systems, and so on.

Costs for services This area includes many of the costs which cannot be assigned, such as project management, design services, planning, etc. These are regarded as one-off costs.

Operating costs Key elements of the operating costs for countermeasures are, for example, the efficiency of the energy conversion of energy storage devices and the operation of auxiliary equipment (pumps, cooling units, fans, control units, etc.).

Maintenance costs It is likely that routine maintenance occurs, particularly in a demonstration project. The annual maintenance costs are seen as a portion of the initial installation costs. There may also be some costs for disposal/shutdown (positive or negative).

6.6 Example of an application: planning an active filter UPCS project

The following guidelines for the design of the UPCS, presented in section 6.4, to deal with perturbations due to harmonics and also due to voltage sags or flicker are explained by using examples of specific cases. The basis of the design instructions are simulation tools, which are prepared using a simulation model [6]. Using the simulation parameters and simulation results, a method has been developed which enables the UPCS to be designed for any load constellations and to meet different requirements regarding voltage quality.

An example of an application is then shown of how the use of the UPCS is to be dealt with as part of network planning.

6.6.1 Designing the UPCS

The connection point for the following examples are defined so that, as shown in Figure 6.27, the converter load under consideration is connected at the connecting point with the simulation model of the UPCS. The system impedance and the busbar voltage U_{SS} which are given for the simulation model are thus the same as those with which the harmonic voltages in particular are calculated.

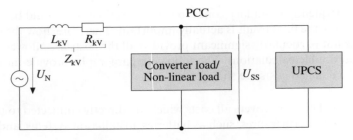

Figure 6.27   Connection of UPCS for simulation

6.6.1.1 Design of the UPCS to compensate for harmonics

Figure 6.28 is a flow chart which shows the functional operation of the calculation algorithm of the design principle.

Figure 6.29 shows a user interface adapted to this calculation algorithm and used for the application of the design principle.

The algorithm shown on this interface can be divided into various successive sections. In the first part, the system parameters for the connecting point of the filter are entered. The voltage and short-circuit power of the primary system, the transformer output and the relative short-circuit voltages u_r and u_x of the

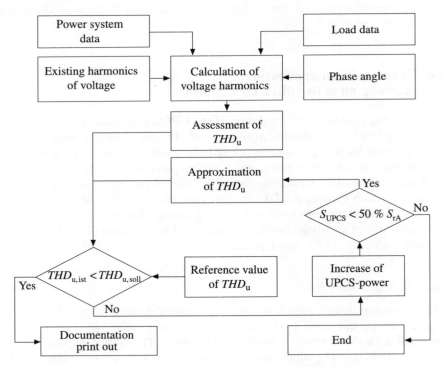

Figure 6.28   Flow chart for determination of UPCS-power (harmonics)

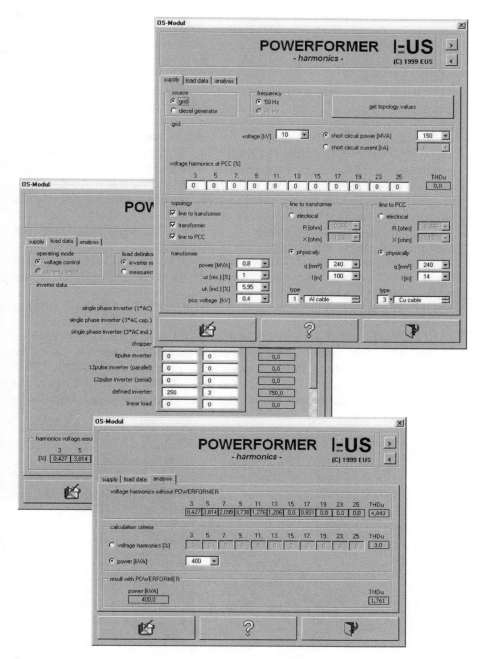

Figure 6.29   Operating window of program to design UPCS (harmonics)

transformer are required for this purpose. The short-circuit power for the connecting point of the filter is calculated from this data.

In the second part, the data of the harmonics source are entered. To do this, a preloading of the industrial network by the primary system can first be considered.

The harmonics created in the industrial network itself are determined by a combination of the converters connected in the network. For seven different types of converter, three power classes, each with a variable number of units, are included. It is also possible to add converters which cannot be assigned to one of these types. However, to do this the relative harmonic currents of the converter must be known. Linear loads which generate no harmonics can also be detected, in order to assess the total loading of the transformer. Five different values for a required THD_u can then be entered. The UPCS power is output relative to the transformer power on the one hand and in kilowatts on the other.

The compensation characteristics of the UPCS for single-phase converter loads are shown in the following as an example of the simulation series on which the design procedure is based. Connection of these converter loads according to Figure 6.27 is assumed. These converter loads are connected to a low voltage end short-circuit power of 25 MVA.

In this way, the range from a relatively-rigid to a weak industrial network is completely covered.

The simulation series proceed with a gradual increase in the UPCS power of 5% up to 50% of the converter design power in each case. Figure 6.30 shows the absolute reduction of the third harmonic as a percentage for various UPCS powers. In this case it must be taken into account that the amplitudes of the harmonics in the networks with a higher short-circuit power such as at

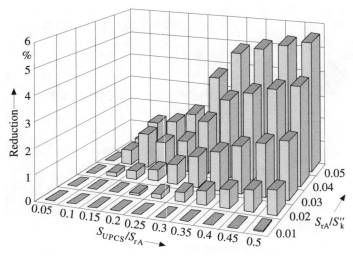

Figure 6.30   Reduction of 3rd harmonic related to power of UPCS S_{UPCS}, power of consumer S_{rA} and short-circuit power S''_k

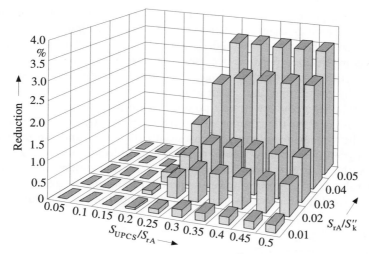

Figure 6.31   Reduction of 11th harmonic related to power of UPCS S_{UPCS}, power of consumer S_{rA} and short-circuit power S''_k

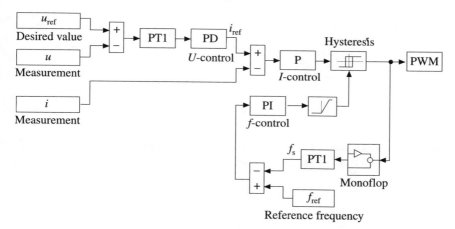

Figure 6.32   Block diagram of UPCS-control

$S_{rA}/S''_k = 0.01$ are generally very low and at $S_{rA}/S''_k = 0.05$ are very high. Characteristic of the course of the absolute reduction for single-phase converter loads is the approximately-linear rise between the power ratios $S_{UPCS}/S_{rA} = 0.2$ and $S_{UPCS}/S_{rA} = 0.35$. The compensation of the third harmonic can be substantially improved in this range by increasing the UPCS power. Further increases in the UPCS power lead to saturation of the reduction.

As the recording of the reduction of the eleventh harmonic in Figure 6.31 shows, the compensation behaviour appears different at harmonics of a higher order.

The compensation effect at higher-frequency harmonics is already stronger at a lower UPCS power, because a differentiating portion of the PID controller of the UPCS (see block diagram of the control system in Figure 6.32) intervenes more efficiently, because of the higher voltage change speed at high frequencies, and less compensation power is necessary. Because of the strong intervention of the D controller, the saturation range of the absolute reduction is also achieved substantially faster, and thus saturation occurs at a filter power of approximately 30% of the converter rated power. Furthermore, a lower compensation current is needed for the compensation of higher-frequency voltage deviations because the system impedance increases with frequency and the higher-order harmonics can thus be significantly reduced even at lower filter ratings.

Where the UPCS is used to compensate for harmonics of single-phase converter loads, the UPCS power required for a specific reduction of the harmonic amplitudes depends heavily on the amplitude of the third and fifth

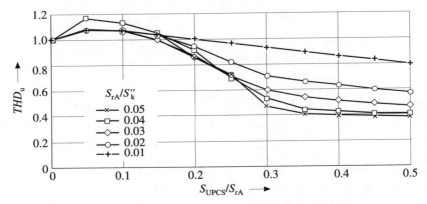

Figure 6.33   Reduction of THD_U of single-phase rectifier harmonic related to power of UPCS S_{UPCS}, power of consumer S_{rA} and short-circuit power S''_k

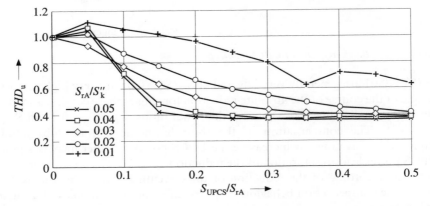

Figure 6.34   Reduction of THD_U of six-pulse converter related to power of UPCS S_{UPCS}, power of consumer S_{rA} and short-circuit power S''_k

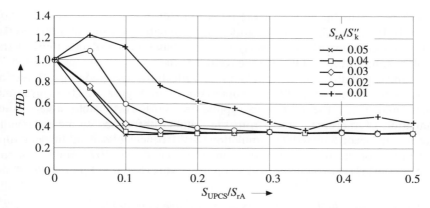

Figure 6.35   *Reduction of THD_U of twelve-pulse converter related to power of UPCS S_{UPCS}, power of consumer S_{rA} and short-circuit power S''_k*

harmonics. In order to reduce the amplitudes of these harmonic orders significantly, a UPCS power of approximately 30% of the converter rated power is necessary (see Figure 6.31), whereby the reduction in the higher order harmonics is substantially greater at this filter rating.

The evaluations of the simulations with regard to total harmonic distortion THD_u for the various converter loads are shown in Figures 6.33 to 6.35. The reduction factors of the THD_u are given in relative quantities so that the absolute reduction of the THD_u above the relative value and that of the original harmonic distortion can be determined without the active filter. The initial increase of the relative THD_u values at low UPCS powers is due to the fact that at very high disturbance levels and relatively low UPCS power the limiting of the compensation current takes place very frequently due to the finite energy content of the intermediate circuit. The consequence of this is, that due to the very incomplete compensation of the harmonics at very small UPCS powers, harmonics of any order can also be generated.

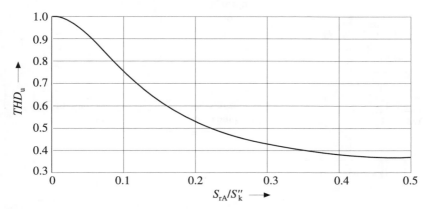

Figure 6.36   *Approximation of THD_U (see Figure 6.27) ($S_{rA}/S''_k = 0.03$) with cubic spline*

With regard to the course of the relative THD_u for the S_{rA}/S''_k ratio of 0.01 shown in Figure 6.35, it must be added that the absolute value of THD_u in the relatively rigid network is extremely low and thus very low fluctuations of the absolute value strongly influence the relative value.

For the calculations of the UPCS power with the design guidelines, the series of measurements of the THD_u course (shown in Figures 6.33 to 6.35) are approximated with the aid of cubic spline functions. This enables a very exact simulation of the curve trace from the data of the series of measurements determined according to the simulations and the calculated additional support points of the functions. The approximation of the THD_u course for the $S_{rA}/S''_k = 0.03$ power ratio recorded in Figure 6.34 is shown in Figure 6.36.

6.6.1.2　*Design of the UPCS with regard to voltage sags and flicker*

In the following, a procedure for designing the UPCS with respect to voltage sags and flicker is presented, similar to the procedure given in 6.6.1.1, and is explained using examples. Figure 6.37 is a flow chart of the algorithm on which the design guidelines are based.

The relative voltage changes due to the load are first determined as a percentage using the system and load data. The A_{st} flicker disturbance factor is

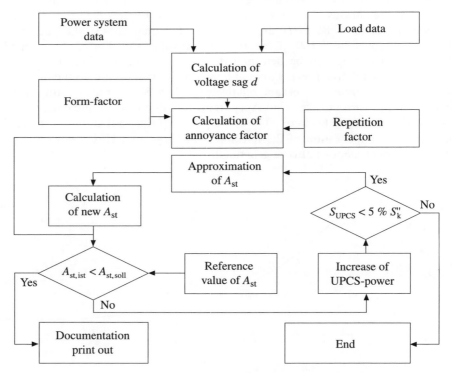

Figure 6.37　Flow chart of program for the design of UPCS (voltage sags and flicker)

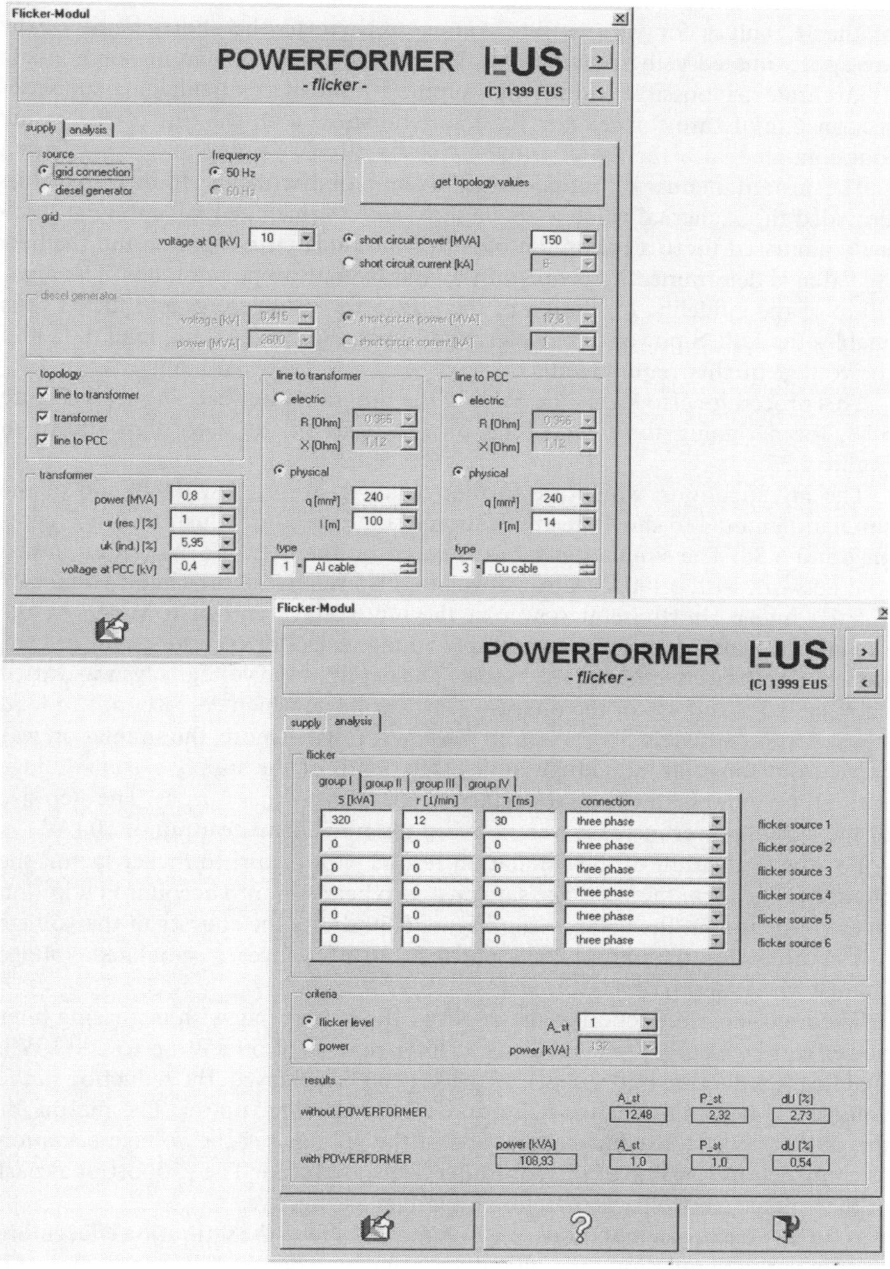

Figure 6.38   Operating window of program to design UPCS (voltage sag and flicker)

determined from the specification of the repetition rates and form factor of the voltage changes (see also Chapter 3). The courses of the reduction factors of the A_{st} values for various filter ratings relative to the short-circuit power are approximated using polynomials. To make the approximation polynomials as accurate as possible, additional support points, as already explained in section 6.6.1.1, are formed for the approximation with the aid of the spline function.

The instantaneous A_{st} actual value is compared with an A_{st} desired value. Provided the A_{st} actual value is above the desired value, the filter rating is gradually increased up to a maximum of 5% of the short-circuit power and the new A_{st} value is determined by means of the approximation polynomials. The algorithm of the guideline for design of the UPCS for voltage sags and flicker thus enables the UPCS power to be determined from the system data, load data and the voltage quality requirements.

This procedure also forms the basis of the program interface, shown in Figure 6.38, for designing the filter rating on the basis of the algorithm shown in Figure 6.37.

The investigations, which are further detailed, are used with the aid of the simulation model to show the application of the procedure illustrated in Figures 6.37 and 6.38. The simulation series is based on the network data of an industrial network with a 400 kVA transformer with a relative short-circuit voltage of $u_k = 4\%$ and a short-circuit power at the low voltage end of 6 MVA. At the various simulations, square-wave supply voltage sags (form factor equal to 1) of various depths were input to the model. The depth of the voltage sags was varied between 0.5% and 4% of the phase-to-earth voltage, which corresponds to load steps of approximately 50 kVA up to 400 kVA. Furthermore, the simulation was carried out for each individual voltage sag depth of the supply system voltage with UPCS powers of 0% up to a 100% of the transformer output. The step size of the UPCS power increase was 12.5% of the transformer output or 50 kW.

For the evaluation of the simulation results with regard to flicker factor, the change in the form factor of the sag must also be taken into account in addition to the reduction in the voltage sag. Figure 6.39 shows the courses of the voltage r.m.s. values for different UPCS powers and also gives a simulated voltage change with a depth of 4%.

The non-linear reduction in the depth of the voltage sag with increasing filter power can be clearly seen. Whereas at filter powers of 50 kW up to 200 kW a significant reduction in the depth of the sag can be achieved, the reduction in the voltage sag begins to change to saturation at high filter power. The reason for this is that with increasing stabilisation of the voltage sag the voltage deviation as a correction variable of the controller is also reduced. The proportional part of the voltage controller is therefore no longer as effective.

A further series of simulations (see Figure 6.40) shows the saturation effect more clearly. In this simulation series, a voltage sag with a depth of 1% was stipulated. With a filter power of 50 kW, corresponding to approximately 12.5% of the transformer output, the depth of the sag can already be reduced by at least a

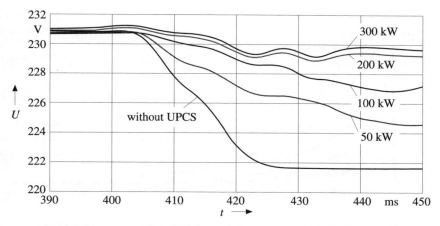

Figure 6.39 Time course of RMS values of voltage in case of voltage sag of 4% with and without application of UPCS (different rating)

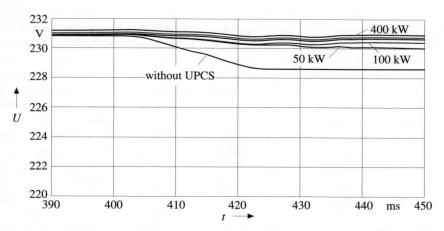

Figure 6.40 Time course of RMS values of voltage in case of voltage sag of 1% with and without application of UPCS (different rating)

half. Doubling the filter rating to 100 kW reduces the voltage sag to only one third of the original depth.

The given rating of the filter and the compensation effect are very heavily dependent on the depth of the relative voltage changes. The greater the disturbance, the higher the compensation power. This means that, at low short-circuit powers, not only is the action of the compensation current more effective but also a higher output is provided by the filter because of the reducing short-circuit power with increasing magnitude of the disturbances.

Figure 6.41 shows the course of the instantaneous flicker factor P_f over a time window of three seconds. The voltage r.m.s. value course on which this

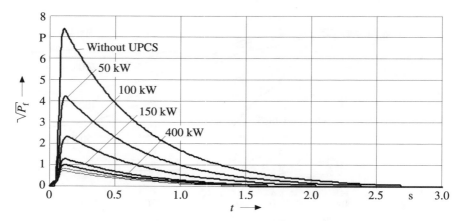

Figure 6.41   Time course of flicker-factor P_f for voltage sag 4% with and without application of UPCS (different rating – see Figure 6.39)

evaluation is based at the beginning contains the voltage changes recorded in Figure 6.39 where $d = 4\%$, whereas the constant voltage values were used for the remainder of the time. In this case, it is only the evaluation of the flicker disturbance factor of a single voltage change during the time frame of three seconds under consideration. Without the filter the relative voltage change d is approximately 4%. The slow decay of the disturbance factor down to zero takes place during the flicker after-effect, which has the effect of a summation of the disturbance factor where voltage changes follow in close succession.

The maximum value of the instantaneous disturbance factor can be assumed to be flicker-determining as an initial approximation [4]. For this reason, the reduction of flicker using the active filter with the instantaneous flicker disturbance factors is evaluated. A strong reduction in the flicker effect can be seen with increasing filter power up to approximately 100 kW. Above this power, the course of the flicker disturbance factor increasingly approaches a minimum. With a filter rating of 100 kW, which in this example is about 25% of the transformer output, the flicker disturbance factor P_f, which completely without a filter is at approximately 3.7 distinctly above the perceptibility threshold, can be reduced to a value of approximately 1.

In Figure 6.42 the evaluation of the instantaneous flicker disturbance factor P_f is similarly shown for the voltage r.m.s. value courses, illustrated in Figure 6.40, with a maximum relative voltage change of 1%. As is expected, the disturbance factor without the filter is clearly lower than for the evaluation for the relative voltage change of $d = 4\%$ shown in Figure 6.41. The slightly stepped curve traces result from the calculation inaccuracy of the evaluation sequence at very low disturbance factors. From a filter rating of 100 kW, or corresponding to 25% of the transformer output, no further substantial improvement in the flicker disturbance factor can be achieved, because the voltage deviation is still very low compared to the set value generated by the filter.

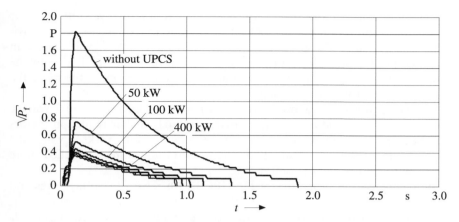

Figure 6.42   Time course of flicker-factor P_f for voltage sag 1% with and without application of UPCS (different rating from Figure 6.40)

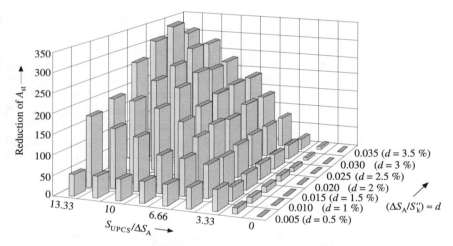

Figure 6.43   Reduction of annoyance factor A_{st} for different voltage sags and different ratios of UPCS-load S_{UPCS} to load change ΔS_A

Reduction factors for the flicker disturbance factor A_{st} are determined on the basis of the evaluations of the instantaneous flicker disturbance factors. Figure 6.43 shows these reduction factors relative to the flicker disturbance factor A_{st} which was achieved by using a filter of different ratings. These are shown depending on the percentage filter power, relative to the load change ΔS_A, and the relative voltage change d which results from the ratio of load change ΔS_A to short-circuit power S''_k. In this case, ΔS_A refers only to load fluctuations and not to the connected load. At a relative voltage change of 0.5%, a saturation of the reduction factor at approximately 50 can be seen at approximately 3.3-times filter power relative to ΔS_A. In this case, the voltage deviation is already so slight that the filter can no longer correct it. It is different at a relative voltage sag of 3% where no saturation is detectable. The reduction factor increases approximately linearly up to 13.3-times the filter power relative to ΔS_A. The increase in the reduction factor with an increasing relative voltage change is due to the reactivity of the PID controller, whose action becomes stronger the more the voltage deviates from the set voltage.

In weak networks with a low short-circuit power in which the relative voltage changes are correspondingly greater, correspondingly-higher reduction factors can be achieved with increasing filter power than in rigid networks in which the relative voltage change is lower and where saturation point is reached more quickly with increasing filter power.

The compensation properties of the UPCS with regard to voltage sags and flicker are thus determined by the depth and shape of the relative voltage sags and short-circuit or transformer output of the industrial connection. The A_{lt} flicker values can be determined from the A_{st} values in that the frequencies of the voltage changes in a 10-minute interval are extrapolated to a time period of two hours.

To determine the effect of the form factor on the reduction factors of the A_{st} values, a ramp-function voltage r.m.s. value is considered in the following and the reduction factors are compared with those of a square-wave voltage characteristic. The system data on which the simulation is based remains unchanged. The maximum relative voltage change without the filter is approximately 1.5%. Figure 6.44 shows the time characteristic of the voltage r.m.s. values for various filter powers. While the voltage over about 220 ms drops slowly in the form of a ramp, it again increases steeply to rated voltage level in a little over 30 ms.

Figure 6.45 shows the reduction factors of the A_{st} values for the ramp-shaped voltage course from Figure 6.44 at various filter powers relative to the short-

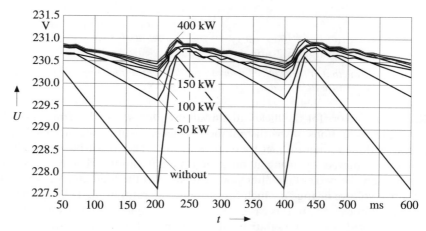

Figure 6.44   Time course of RMS value of voltage in case of ramp-function of voltage, with and without application of UPCS

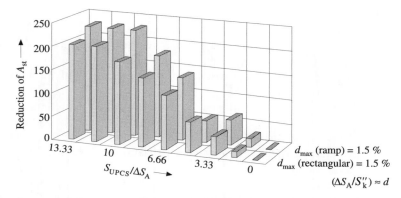

Figure 6.45   Reduction of annoyance factor A_{st} for ramp-function of voltage and different ratios of UPCS-load S_{UPCS} to short-circuit power S''_k

circuit power. Compared with the values for the square wave voltage-change course, it can be seen that these almost exactly coincide so that the compensation properties of the filter in the first place depend on the absolute amount of the relative voltage change *d*.

6.6.2 Example of network planning, taking account of active system filters

Because the system perturbations of large industrial consumers are no longer avoidable, and the voltage quality must be adequate to guarantee sensitive industrial processes, it is necessary to formulate measures here which must be checked with regard to their technical and economic feasibility [4].

The possibility of using active filters to optimise the system connection planning is examined in the following using a large industrial consumer as a realistic example. To do this, the inclusion of an active filter in the system connection planning of a steel works with an 85 MVA d.c. arc furnace is examined. A measurement of the daily load variation of such an arc furnace is recorded in Figure 6.46. The strongly-pulsating power course is due to the statistical formation of the arc. The load impedance fluctuates almost infinitely during idling to near zero in short-circuit.

The problems which arise with regard to the power supply for the arc furnace can be summarised (according to [4]) as follows. Because the random, heavily-varying nature of the load current, voltage changes are caused at the system impedance, which, as shown in Figure 6.47, can be broken down into a longitudinal and transverse voltage drop. Furthermore, d.c. arc furnaces cause harmonics due to the large rectifier stations, which are not dealt with further here, because experience has shown that the periodic voltage changes are designed for the system connection.

Because the inductive component of the system impedance X_N is at least 10 times greater in high voltage systems that the ohmic component R_N, amplitude changes are mainly caused by changes in the reactive power. All other loads

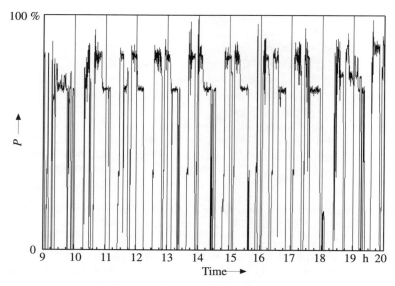

Figure 6.46   Daily load variation of an arc furnace

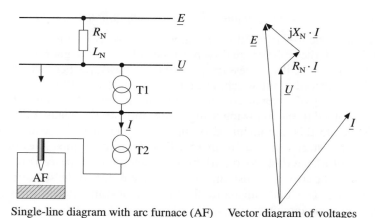

Single-line diagram with arc furnace (AF)   Vector diagram of voltages

Figure 6.47   Connection of an arc furnace and vector diagram of current and voltages

connected at this connection point of the arc furnace are influenced by these voltage changes. Furthermore, the voltage profiles in adjacent network nodes also change, but are not shown in Figure 6.47.

The following investigation is designed to show that by compensating for the system perturbation effects of an arc furnace by using an active system filter it is possible to convert the system connection from a 220 kV busbar to a 110 kV busbar with a lower short-circuit power, and that it is possible to connect other loads. For this purpose, two different connection variants for large industrial customers are considered. The load flow analysis and failure simulations with the various connection variants are used to check the technical feasibility with regard to equipment loading, and the voltage profile is also checked. With the compensation properties of the active filter UPCS with regard to flicker, analysed in section 6.6, the required filter power necessary to reduce the flicker sufficiently at the 110 kV connection to preclude the occurrence of impermissible levels in the public supply system is determined. For this, it is assumed that the filter retains the compensation properties at very large filter powers and short-circuit powers, which is still to be investigated after the development of a method of connecting the UPCS in high voltage systems. Furthermore, load flow analyses are performed with the different connection variants, by means of which the technical feasibility of the individual variants is checked. The aim of these load flow analyses is to check the voltage profiles of the network nodes and the loading of the individual equipment where there are changes in the system configurations. A further investigation will be carried out to show the effect of the various variants on the $(n-1)$ supply integrity. The load flow analyses and the failure simulations are performed using a network calculation program [4].

6.6.2.1 *System connection variants of a large industrial customer*

By way of example, two different variants of the system connection of a large industrial customer are shown. In variant A the customer is connected via a separate 200 MVA transformer to a 220 kV busbar. In variant B the customer is supplied from the 110 kV system from which the public supply system is also fed.

Connection variant A The system connection of the large industrial customer is shown in Figure 6.48. The customer in this case is a steel works which makes steel using a d.c. arc furnace. To avoid impermissible perturbations of this arc furnace acting on the works power supply and the public supply system, this furnace is supplied via a separate 220/110 kV transformer. This transformer, with an output of 200 MVA, is solely responsible for supplying the arc furnace, which emits strong disturbance and has an output of 85 MVA. The arc furnace is thus operated decoupled from the rest of the network. The short-circuit power at the 220 kV busbar is 8.7 GVA. The public supply system and the remaining works supply of the steel works are connected through two further 200 MVA transformers. The short-circuit power in both separate 110 kV busbars amounts to approximately 4.5 GVA for the public supply system and 1.2 GVA for the industrial supply. In the industrial network of the customer, the supply for the d.c. arc furnace is transformed, after an overhead line transmission of approximately 9 km, to medium voltage by means of four 40 MVA transformers. The short-circuit power at the 30 kV busbar amounts to approximately 0.45 GVA.

Because the 220 kV network as a transmission network (see Figure 6.48, location VP) is used in only exceptional cases to supply customers, the measured

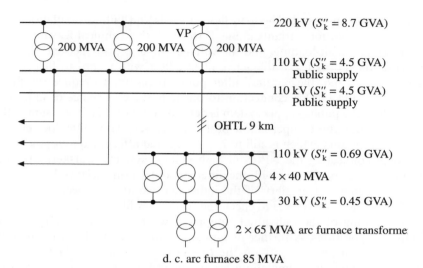

*Figure 6.48   Connection of an arc furnace by separate 220/110 kV transformer
(alternative A)*

*Table 6.2   Maximum values of the flicker disturbance factors for connection of
the arc furnace in accordance with variant A*

Flicker disturbance factors without an active filter	A_{st}	A_{lt}	Permissible?
220 kV connection (VP), $S''_k = 8.7$ GVA	1.2	0.33	Yes
110 kV public supply system, $S''_k = 4.5$ GVA	< 0.8	< 0.3	Yes

flicker factors shown in Table 6.2, which are above the perceptibility limit, cause
no inconvenience. The flicker factor, which in this case occurs at the adjacent
110 kV busbar for the public supply, is at $A_{st} < 0.8$ or $A_{lt} < 0.3$. Experience shows
that this level also causes no problems in the public supply.

Connection variant B   For the variant shown in Figure 6.49, the d.c. arc furnace
is supplied from the 110 kV network. The 110 kV busbars for the public supply
system and the industrial supply of the arc furnace are combined into one
busbar (see Figure 6.49, location VP) with this variant. In contrast to variant
A, this 110 kV busbar is supplied through two 300 MVA transformers from the
220 kV network. The short-circuit power at the 110 kV busbar is then 5.6 GVA.
After the overhead line, the short-circuit power at the 110 kV transfer point
reduces to 2.1 GVA. At the 30 kV busbar of the customer, a short-circuit power
of 0.8 GVA is finally reached.

Table 6.3 shows the flicker disturbance levels of this system connection variant
caused by the arc furnace. The essential facts about these disturbance factors is

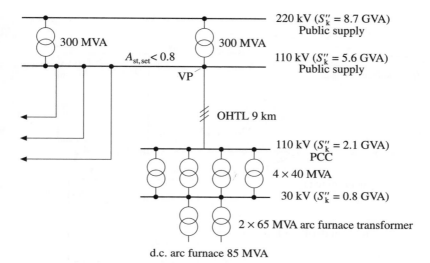

Figure 6.49   *Connection of an arc furnace by separate 110 kV substation or busbar (alternative B)*

Table 6.3   *Maximum values of the flicker disturbance factors for connection of the arc furnace in accordance with variant B*

Flicker disturbance factors without filter	A_{st}	A_{lt}	Permissible?
110 kV busbar (VP) S''_k=5.6 GVA	4.50	1.24	No
110 kV transfer point, S''_k=2.1 GVA	85.32	23.46	No
30 kV busbar of the customer, S''_k=0.8 GVA	1,543.40	424.40	No

that the values at the 110 kV busbar with the public supply system are clearly too high and experience shows that this leads to complaints about flicker. To achieve this system connection variant for industrial operation, countermeasures are necessary with regard to disturbance due to flicker.

To show the possibilities of achieving both variants from a network planning point of view, load flow analyses and failure simulations are first performed. After this, the use of active filters to reduce the disturbing flicker in connection variant B is considered.

Load flow analyses of various system connection variants In the following, the extent to which the A and B system connection variants satisfy the requirements of network planning with regard to the load flow distribution and the $(n - 1)$ criterion is investigated. Analyses of the load flow as well as simulations of the outage of individual equipment are made for the different system connection variants from section 6.6.2.1. The aim of these calculations is to determine the

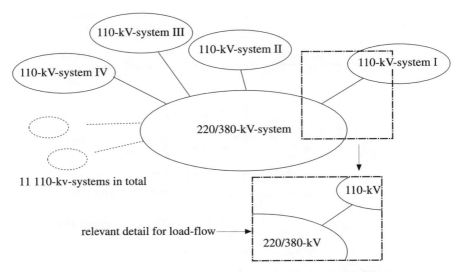

Figure 6.50 *General overview and selected detail of power system topology for load-flow analysis (for details see text)*

voltages in all network nodes and the power flows through all connecting elements between the nodes. From this, the values for the loading of the individual equipment is obtained and the compliance of the limits of the permissible voltage bands.

The calculations are made with the 110 kV network groups and primary 220/380 kV networks, shown in Figure 6.50, by using a network calculation program. The basic data of the loading and the feeds into the individual network nodes are stipulated for a heavy load scenario during winter. To represent the load flow results, a network section is chosen which includes the connection variants of the industrial customer with network areas from the 220/380 kV network and from a 110 kV network group. This limitation is permissible because the changes in the load flow results at a larger electrical distance remain within very narrow limits.

Figure 6.51 shows the result of the load flow analysis of system connection variant A according to Figure 6.48. The power flows at the connecting elements are given in active and reactive power amounts in MW or Mvar. The node voltage amounts at the network nodes are shown in kV. As the calculation has shown, there is no equipment overload or violation of the maximum permissible voltage band of $U_n \pm 10\%$ with this system connection variant. The simulation for verification of the $(n-1)$ criterion produced no critical loading of any equipment of more than 90% of the rated power. From the point of view of network planning, this connection variant can be achieved without reservation according the $(n-1)$ criterion and the load flow.

Figure 6.52 shows the results of the load flow analysis for planning variant B. With this variant too no complications arise with regard to violation of the

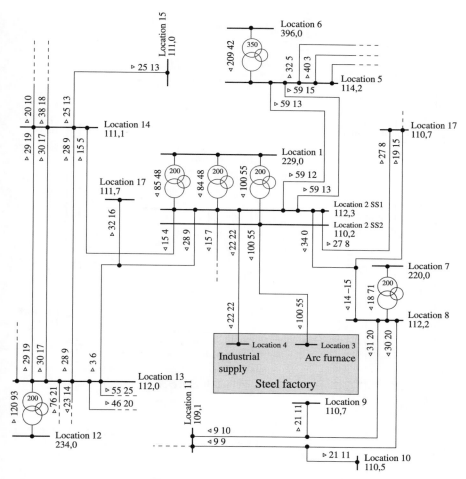

Figure 6.51   *Results of load-flow analysis (alternative A as in Figure 6.48) for peak load*
        figure below 'location':   *Voltage in kV*
        first figure at line: *Active power in MW*
        second figure at line: *Reactive power in Mvar*

Table 6.4   *Reduction factors required for the flicker disturbance factors when considering the various locations of the filter according to Figure 6.53*

Flicker disturbance factors at location	$A_{st,actual}$ (without filter)	A_{st} (with filter)	Required reduction factor
Location 1, S''_k = 5.6 GVA	4.50	< 0.8	Approx. 6
Location 2, S''_k = 2.1 GVA	85.32	< 15.2	Approx. 6
Location 3, S''_k = 0.8 GVA	1,543.40	< 274.40	Approx. 6

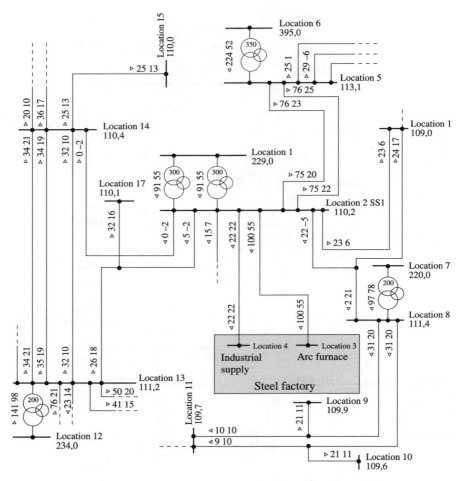

Figure 6.52   Results of load-flow analysis (alternative B as in Figure 6.49) for peak load.
For explanation see Figure 6.51

maximum permissible voltage deviations or an excessively increased equipment loading.

Figure 6.53 shows the result of the failure calculation to check the $(n-1)$ criterion. If a case occurs where one of the two 300 MVA transformers between nodes 1 and 2 (which feed the 110 kV network from the 220 kV network) fails, an equipment loading of approximately 97% occurs at the 200 MVA transformer between nodes 12 and 13. At this loading, a particular monitoring of the transformer in such a scenario is recommended. This system connection variant of the industrial customer can therefore be achieved without problems from the point of view of the load flow and the $(n-1)$ criterion.

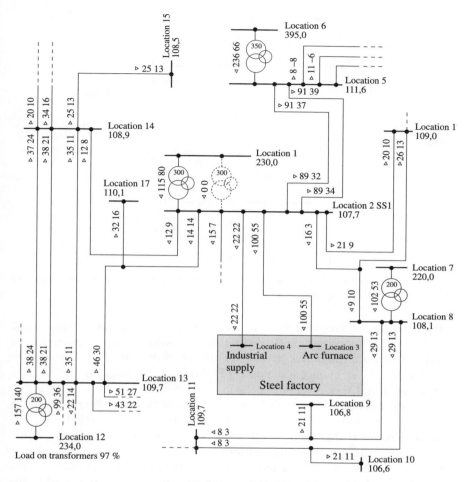

Figure 6.53   *Results of load-flow analysis (alternative B as in Figure 6.49) for peak load*
Outage of one 300-MVA-transformer between 'Knoten 1' and 'Knoten 2'
For explanation see Figure 6.51

Active filters in the system connection planning   After the fundamental consideration of the system connection variants with regard to the load flow analyses, the remedial measures necessary for the reduction of the flicker disturbance effect in connection variant B is checked.

In order to do this, three possible ways of using an active filter are more closely considered. The filter can be connected, as shown in Figure 6.54, directly to the 110 kV busbar of the public supply (location 1, $S''_k = 5.6$ GVA) or it can be connected to the 110 kV customer connection (location 2, $S''_k = 2.1$ GVA) after the 9 km long overhead line, or connected to the 30 kV level within the industrial network (location 3, $S''_k = 0.8$ GVA).

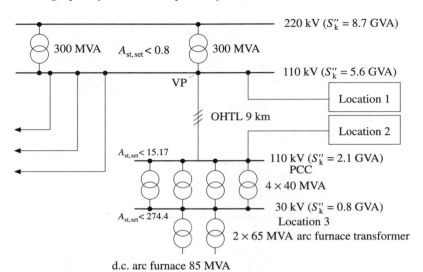

Figure 6.54   Connection of an arc furnace (alternative B as in Figure 6.49), application of active filter at different locations

The disturbance factor at the 110 kV busbar for the public supply where $A_{st} = 4.5$ is clearly too high, so that with this connection variant measures must be taken to reduce the flicker level. To avoid complaints about flicker, the disturbance factors at connection point VP should be reduced to the level of $A_{st} < 0.8$ or $A_{lt} < 0.3$.

This results in three different filter powers, because the short-circuit powers in the connection points of the filter (and thus also the level of the relative voltage changes effective at that point which must be considered for the design of the filter) are different in the various locations.

Because the operating behaviour of the arc furnace is independent of the voltage level at the customer connection, the only difference, with regard to the calculation of the flicker disturbance factor between the 220 kV, the 110 kV and the 30 kV connection, is in the level of the relative voltage changes d caused by the load fluctuations, which are included cubically in the calculation of the flicker disturbance factor. The relative voltage change, and thus also the flicker disturbance factor, in the 110 kV and 30 kV voltage levels can be assessed using Equations (6.7) and (6.8).

$$d_{110\,kV} = d_{220kV} \cdot \left(\frac{S''_{k\,220kV}}{S''_{k\,110\,kV}}\right) \tag{6.7}$$

$$A_{st\,110\,kV} = A_{st\,220\,kV} \cdot \left(\frac{S''_{k\,220kV}}{S''_{k\,110\,kV}}\right)^3 \tag{6.8}$$

The A_{lt} values are calculated in the same way.

The short-circuit power at the 110 kV busbar is determined by a short-circuit

Table 6.5   Absolute and relative filter powers for various locations

Location of filter	S_{filter}/[MVA]	$S_{filter}/\Delta S_A$	S_{filter}/S''_k
Location 1, S''_k = 5.6 GVA	20	0.66	0.0035
Location 2, S''_k = 2.1 GVA	8	0.26	0.0041
Location 3, S''_k = 0.8 GVA	6	0.20	0.0084

current calculation because of the multiple feeding into the 110 kV voltage level. With the flicker disturbance factor at the end of the 110 kV customer connection, which initially appears too high, with a short-circuit power of 2.1 GVA, it must be taken into account that this is not the point of common coupling and therefore the A_{st} disturbance factor must not be reduced to 0.8. The reduction must be chosen so that the level does not exceed 0.8 at the connection point with the public 110 kV network. That would be quickly reached if the A_{st} value at the end of the 110 kV customer connection was 15.2. By analogy with the previous statements on flicker level, the A_{st} values at the 30 kV busbar may reach about 275 without having a disturbing effect on the public 110 kV supply.

To reduce the flicker disturbance factor A_{st} to an acceptable level at the 110 kV busbar, from which the public network is also fed, a reduction factor for the flicker level of about 6 is required at the three locations. The deciding factor for determining the relative voltage change d is the load deviation which, according to Figure 6.39 can, on the one hand, be assumed with a 100% furnace output and a repetition rate of approximately 2 in 10 minutes or, on the other hand, with 20% of furnace output and a repetition rate of 700 in 10 minutes. The frequently-occurring smaller load fluctuations, which for the arc furnace considered here are between 10 MVA and 30 MVA, are usually far more critical with regard to flicker. The filter rating using the design guidelines developed in section 6.6 results in the UPCS powers shown in Table 6.5 for the different locations.

From this table it can be seen that the filter power must be greater with an increase in the electrical distance from the cause of the disturbance, because of the increasing short-circuit power. The compensation power supplied by the filter, which depends on the depth of the relative voltage change and thus on the $\Delta S_A/S''_k$ ratio, increases with an increase in the control deviation which is determined by this voltage change d.

6.6.2.2 Optimisation of the location of active filters

The effectiveness of an active network filter is largely determined by the location. The compensation properties of the UPCS depend, for example, on the magnitude of the disturbance level which occurs, and on the short-circuit power at the connection point of the filter. Therefore the filter should be fitted as close as possible to the disturbing operation or disturbing system. In cases where there are just a few minor complaints about disturbances such as flicker or harmonics, it is possible to consider a decentralised compensation compared with a central compensation at the point of disturbance. If the total filter power required for decentralised compensation is lower, the advantages and disadvantages of the possibilities must be weighed against the background of future load and network developments. In this respect, active filters offer the great advantage of mobility, so that changes in the operating location itself do not limit the application possibilities of the filter.

6.6.3 Assessment of active network filters from the point of view of network planning

For large industrial customers, the power supply represents an important competitive and location factor for a claim on the market. The demands placed by some industrial customers on the energy supplier is likely to increase, in the future, due to competition. To meet these changed demands, it is the task of power supply companies to carry out a requirement analysis with the customer and work out an appropriately-tailored service which is then also reflected in the price structure.

The local reduction of the system perturbations of an industrial operation by the use of active network filters can also be part of such a service. By using active filters in the network planning, the number of possible planning variants, and thus the flexibility, can be increased. These variants then differ with regard to costs, the possible realisation time period and the resulting power quality for the customer, so that with regard to these criteria a customer-specific optimum solution can be found. The failure costs and failure secondary costs due to an unmatched power quality which, for example, could be reduced by an active filter, are of particular consideration here. The advantages of making the system connection of a customer flexible in this way can be used particularly to satisfy the customer needs. This can lead to cost savings for both the customer and power supply company and competitive advantages for both sides. A further advantage which the active network filter brings to network planning is in the supply of extremely voltage-sensitive loads. These can be brought into use in the course of modernisation, with a voltage quality locally matched to the requirements, so that expensive network expansion and conversion planning can be avoided.

6.7 References

1   MOHAN, N., UNDELAND. T., and ROBBINS. T.: 'Power Electronics', (John Wiley & Sons, 1995, 2nd edn.)
2   RATERING-SCHNITZLER, B., and KRIEGLER, U.: 'Versorgungssicherheit und Spannungsqualität durch UPCS (Supply integrity and

voltage quality through UPCS)' Unified Power Conditioning System, VDI-Fachtagung 'USV and Sicherheitsstromversorgung III'. Leipzig, November 1996

3   RATERING-SCHNITZLER, B.: 'Einsatz eines Schwungmassenspeichers zur Überbrückung von Spannungseinbrücken und kurzfristigen Versorgungsunterbrechungen (Use of a gyrating mass flywheel for bridging voltage sags and short-term supply interruptions)' VDI-Fachtagung Energiespeicherung für elektrische Netze. Gelsenkirchen, November 1998

4   BRIEST, R., and DARRELMANN, H.: 'Alternative Power Storages for UPS-Systems' Conference Proceedings 'European Power Quality 97', ZM Communications GmbH, 1997, pp. 371–372

5   APELT, O., HOPPE, W., HANDSCHIN, E., and STEPHANBLOME, T.: 'LimSoft—Ein innovativer leistungselektronischer Stoßkurzschlußstrombegrenzer (An innovative power electronic sudden short-circuit current limiter)' *Elektrizitätswirtschaft* 96, 1997, vol. 26, pp. 1599–1603

6   SCHROEDER, M.: 'Einsatz und Auslegung aktiver Filter zur Netzrückwirkungskompensation (Use and design of active filters for system perturbation compensation)' *Technischer Bericht der EUS GmbH*, Gelsenkirchen, 1997

7   DANEK, H.D.: 'Einflußvon Kondensatoren auf die Netzqualitat' (Influence of capacitors on voltage quality). ABB Kondensatoren GmbH, Report P109 E104

8   BLUME, D., DANEK, H.D., SCHLABBACH, J., and STEPHANBLOME, T.: 'Messung und Bewertung von Netzrückwirkungen' (Measurement and assessment of voltage quality). Haus der Technik, Essen, 1996, Report E-10-222-073-6

Chapter 7

Notes on practical procedures

7.1 Survey of voltage quality (harmonics) in medium voltage networks

Task For medium and low voltage networks of public and industrial electrical supply systems a survey of existing voltage harmonics and also, under certain circumstances, their change over a period of a year or development over several years is of interest. The following procedure is recommended:

Procurement of network data
Analysis of network plan
– Voltage level, cables and overhead lines, supply voltage levels.
– Network data, short-circuit power.

Analysis of consumer structure according to voltage levels
– Low voltage network
    Residential areas, rural areas, trade areas such as offices, business houses and department stores in town centres, special consumers.
– Medium voltage network
    Municipal supply or rural supply with, and without, industrial or trade loads, industrial supply, feeds for converters for traction supply, any self-generating systems (wind energy or photovoltaic).

Aspects of measurement
Specification of a measuring program
– Time duration one week (weekdays and weekends).
– Annual course, summer and winter measurements, low load and peak load season.

Measurement of voltage harmonics
– Measurement of current harmonics where there are large harmonics generators or consumer groups.

Repetition of measurements and evaluations
–   Equal or similar load conditions for measurements over various years.
–   Measurements at different seasons of the year.

Evaluation and assessment
–   Evaluation of measurements with comparison of different loads, days of the week etc.
–   Comparison of workdays and weekends.
–   Action required if the compatibility levels are exceeded.

Example Some measurements are given in the following by way of example. The measurement results are not shown standardised, because different systems were used for the measurements.

The results of further systematic measurements are given in [1].

The pronounced characteristic of the fifth voltage harmonic due to the increased utilisation of consumer electronic equipment (television sets) in the evenings and at weekends can be seen in Figure 7.1. The total increase in voltage levels at the weekend is also due to the reduced network load and the associated lower attenuation.

Figure 7.2 shows the steep rise in the fifth voltage harmonic in a developing area over a period of five years, caused by the increase in domestic loads due to consumer electronic equipment.

Figure 7.3 shows the course of the harmonic voltages of the 5th order (Figure 7.3a) and 13th order (Figure 7.3b) for a 30 kV industrial network with a predominant load through twelve-pulse converters. It can be seen that the course of

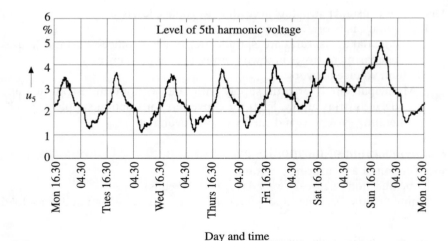

Figure 7.1   Time course of 5th harmonic of voltage in a 10 kV system during one week in July; urban area, $P_{max} = 8.7$ MW

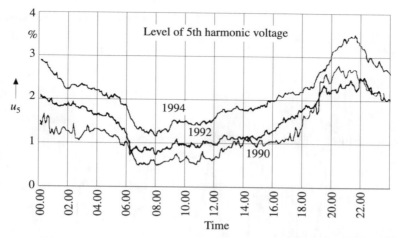

Figure 7.2   *Time course of 5th harmonic of voltage in a 10 kV system during one working day in September for various years; urban area with small industrial consumers 1990: P_{max} = 4 MW; 1992: P_{max} = 7.1 MW; 1994: P_{max} = 16 MW*

the 5th harmonic is caused only slightly by the converter systems, but instead is predominantly due to the medium voltage networks connected to the primary 110 kV network. The course of the 13th harmonic voltage, on the other hand, is determined by the converters. It can also be seen from the course of the harmonic voltage that with the supply current remaining almost equal to the harmonic voltage on Friday it is substantially lower than on the previous days. This is clearly due to the change in the network configuration, which causes a network resonance which was previously present to be shifted, or the network impedance to be reduced.

Figure 7.4 shows the course of the fundamental component active power and also the harmonic voltages of the orders 5, 7, 11 and 13 in a public supply system with a connected industrial operation whose main load is represented by a twelve-pulse converter where $P = 4.1$ MW [2] (the network arrangement is shown in Figure 7.5). In this case the slight influence of the non-characteristic harmonics ($h = 5, 7$) and also the dominating influence of the characterising harmonics ($h = 11, 13$) can be clearly seen.

7.2 Connection of harmonics generators, high-load consumers

Task The connection of high-load, harmonics-generating consumers, such as converter motors, battery storage systems and converters for industrial heating equipment cannot be assessed from the emitted interference. Instead it is necessary to perform network analyses, network measurements and, if necessary, calculations to check the permissible operation of systems. The basic procedure shown in the following is further explained by using a medium-frequency induction furnace as an example.

a)

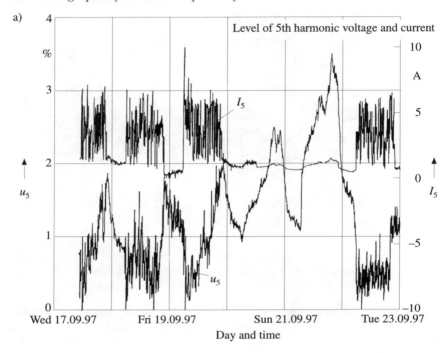

b)

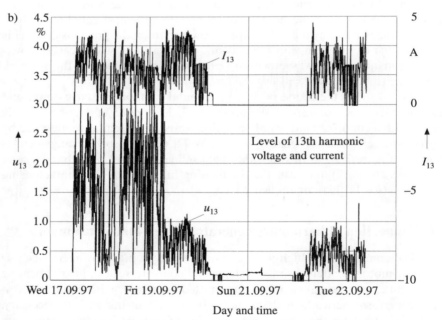

Figure 7.3   a)   Time course of 5th harmonic of voltage and current in a 30 kV system.
Measuring period 10 a.m. Wednesday until 10 a.m. Tuesday
b)   Time course of 13th harmonic of voltage and current in a 30 kV system

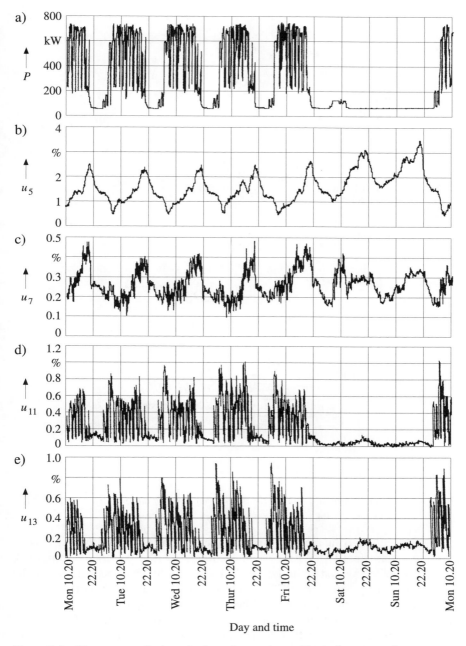

Figure 7.4   Time course of selected voltage harmonics and basic frequency of active power in a 10 kV system with rectifier load P = 4.1 MW, measured for one week

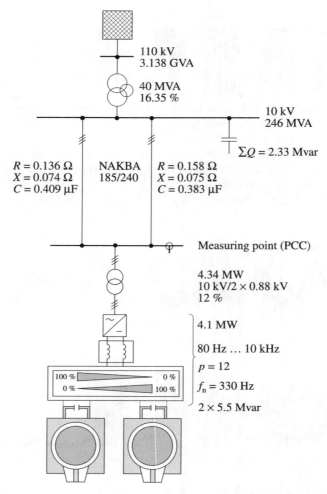

110 kV
3.138 GVA

40 MVA
16.35 %

10 kV
246 MVA

$\Sigma Q = 2.33$ Mvar

$R = 0.136\ \Omega$
$X = 0.074\ \Omega$
$C = 0.409\ \mu\mathrm{F}$

NAKBA
185/240

$R = 0.158\ \Omega$
$X = 0.075\ \Omega$
$C = 0.383\ \mu\mathrm{F}$

Measuring point (PCC)

4.34 MW
10 kV/2 × 0.88 kV
12 %

4.1 MW

80 Hz … 10 kHz

$p = 12$

$f_\mathrm{n} = 330$ Hz

2 × 5.5 Mvar

100 %   0 %
0 %   100 %

*Figure 7.5   Power system diagram for the connection of a medium frequency converter for
inductive melting*

The connection of a medium-frequency converter ($S_\mathrm{r} = 4.84$ MVA; twelve-pulse) to a municipal 10 kV network in accordance with Figure 7.5 is examined. The 10 kV switching system in the 110/10 kV substation is to be regarded as the point of common coupling (PCC), because it is only at that point that other consumers are supplied from the public system. The supplied 10 kV network is operated as a radial network. Significant changes to the switching state are not possible, in particular a further supply of the 10 kV network from a different 110/10 kV substation is not possible.

Network data   The network data in Figure 7.5 is necessary for the assessment of the connection of the converter.

According to the manufacturer's data, the harmonic currents of the converter during rated operation are as follows:

$$I_5 = 5.03 \text{ A}; I_7 = 3.19 \text{ A}; I_{11} = 13.92 \text{ A}; I_{13} = 8.61 \text{ A};$$

$$I_{17} = 0.34 \text{ A}; I_{19} = 0.31 \text{ A}; I_{23} = 2.43 \text{ A}; I_{25} = 2.46 \text{ A}.$$

Aspects of measurement The load profile of the industrial operation was recorded over one week to establish a suitable assessment time frame. Figures 7.6 and 7.7 show a periodicity of a daily and weekly operating pattern from the time-course of the harmonic voltages at the PCC, using the eleventh voltage harmonic, together with the time-course of the fundamental component active power, as an example. This is clearly a two-shift operation.

The harmonic voltages and currents were measured under defined operating

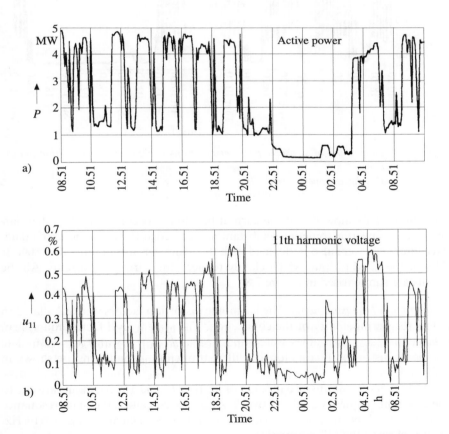

Figure 7.6   Time course of measured parameters during one day; operating conditions c)
as in Figure 7.8
a)   basic frequency of active power
b)   voltage harmonic of order 11

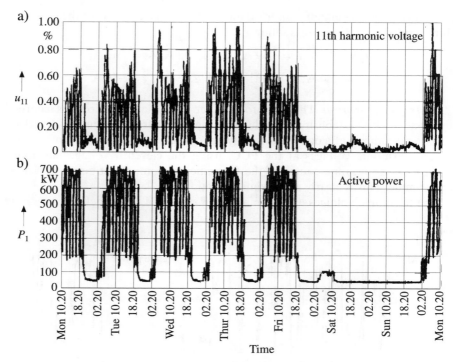

Figure 7.7   Time course of measured parameters for one week; operating condition c) as in
Figure 7.8
a) voltage harmonic of order 11
b) basic frequency of active power of one phase

conditions. This showed that the current harmonics occurring at the rated output of the converter sometimes substantially exceeded the manufacturer's data.
Measurements were also performed with the converter output limited to 80% or
73% of the rated output, which clearly reduced the harmonic currents. All the
results are summarised in Figure 7.8.

Assessment of measurements The harmonic distortion factors were calculated to
assess the permissibility of the connection. The system level factor and system
connection factor variables were stipulated as $k_{NMV} = 0.4$ and $k_A = 0.16$. The
harmonic distortion factors for the different operating conditions are shown in
Table 7.1.

The high harmonic disturbance factor for the 11th and 13th harmonic are due to
the resonant frequency of the network. At the given values, the main resonance
of the network at the point of common coupling is calculated as $f_{res} \approx 514$ Hz,
i.e. it is close to the 11th harmonic.

Summary and conclusion On the basis of the harmonic disturbance factors B_h
according to Table 7.1, the assessment shows that unrestricted operation of the

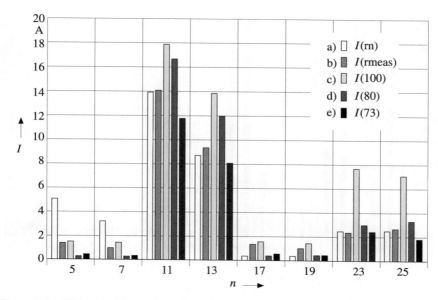

Figure 7.8 95% probability of significant harmonics for different operating conditions
    a) rated power according to manufacturer (rn = rated nominal)
    b) rated power as per installation (4.84 MVA, 4.1 MW)
    c) maximal loading of induction furnace
    d) 80% of rated power as per installation
    e) 73% of rated power as per installation

Table 7.1 Harmonic disturbance factors B_h (95% frequency) of the converter,
with operating conditions according to Figure 7.8

Operating condition Harmonic order	a)	b)	c)	d)	e)
5	0.03	0.008	0.009	0.001	0.002
7	0.031	0.009	0.014	0.002	0.003
11	0.36	0.365	0.462	0.432	0.302
13	0.417	0.448	0.668	0.576	0.395
17	0.09	0.344	0.419	0.097	0.142
19	0.052	0.166	0.242	0.119	0.076
23	0.127	0.122	0.395	0.151	0.125
25	0.097	0.104	0.276	0.129	0.071

system is not permissible because the maximum harmonic disturbance factor
(B_{13}) under all operating conditions or converter settings is above the load-
proportional permissible value ($B_h > k_A \times k_{NMV}$). Where the converter output is
limited to 73% of the rated output, the maximum harmonic disturbance factor
is, however, below the value permissible for the network level ($B_h > k_{NMV}$).

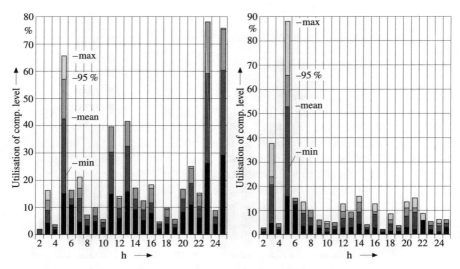

*Figure 7.9   Statistical parameters for utilisation of compatibility level (harmonic voltage)
at PCC; operating condition c) as in Figure 7.8
Measuring period 6 a.m. Monday until 7 p.m. Friday*

Operation of the plant can thus be approved, provided no further significant harmonics generators are connected at the same system connection point (PCC), or these do not completely take up the harmonics level assigned to them. This applies in the present case. It can be concluded from the assessment that the limitation of the converter output to 73% of the rated output was integrated into the control concept. In this special case, there was no significant effect on the operating process, i.e. it was not necessary to extend the smelting or pouring time of a batch despite the output limitation.

Figure 7.9 shows the statistical parameters of the voltage harmonics of orders 2 to 25 for operation of the converter, limited to $0.73 \times P_r$ over the Monday to Friday measuring period.

7.3 Determining the reference values for planning calculations in a ring-cable network

7.3.1 Measurements in 35 kV ring-cable network

Task In an extended cable network it was necessary to examine whether the relocation of a capacitor bank, which was required to maintain the voltage, could also be carried out with regard to aspects of voltage quality without causing impermissible harmonic voltages in the affected network. The measurements were used to determine the harmonic levels in the network.

Data procurement Using the network plans, the measuring points were located so that various network groups could be measured. The measuring channels already available were also used to measure the supply behaviour of the main converters with regard to harmonic currents. An overview of the network, showing the relevant measuring points, is given in Figure 7.10 [3].

Measurement results/assessment   The essential results of the measurements of the harmonics are summarised in the values for the total harmonic distortion (*THD*). These values are registered at the corresponding measuring points in Figure 7.10. The harmonic voltages and currents were measured. The measured values were stored in the form of values averaged over one minute. Figure 7.11 shows an example of the time course of the *THD* at a chosen measuring point. The industrial network did not have a regular load curve at any measuring point.

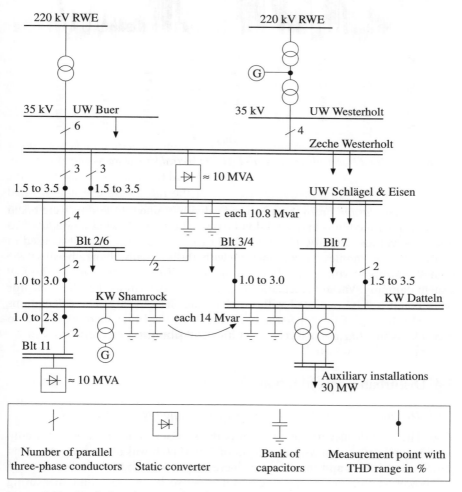

Figure 7.10   Single-line diagram of medium voltage system

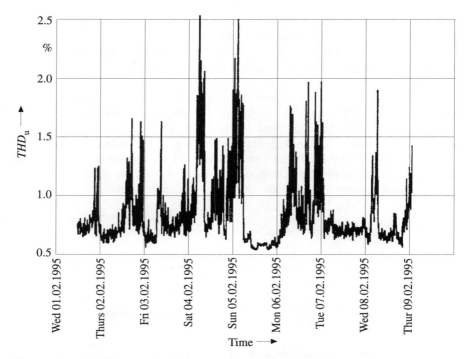

Figure 7.11   Total Harmonic Distortion (THD) measured for one week

Summary/conclusion From the measured values obtained, time slots were taken in the heavy-, medium- and weak-load cases. The heavy-load case was used to assess the voltage stability. The weak-load case, on the other hand, was used to analyse the harmonics in conjunction with the capacitor bank, because the highest harmonic voltage levels occur in the weak-load case due to the low system damping. Variant calculations were used to determine that the capacitor bank could not be relocated without changing the capacity. The calculations enabled a tuning of the capacitor bank which ensured that there would be no excessively-high harmonic levels under any load situations.

7.4 Disturbance investigation

7.4.1 Disturbance analysis harmonics in power station service network I

Task The case dealt with here concerns the station service network of a conventional power station with an output of 520 MW. It was observed over a long period that defects and disturbances increasingly occurred in various areas of the power station on parts of computer networks, copiers and measuring instruments.

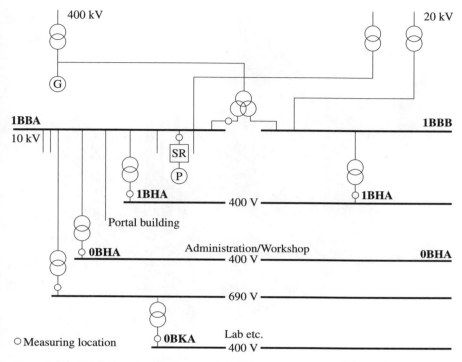

Figure 7.12   Single-line diagram of low voltage system in a power station

Data procurement Figure 7.12 shows the system diagram and the technical data.

Aspects of measurement To examine the effects, a long-term measurement of the harmonics and flicker levels was performed, supplemented by measurements of the time signal. Readings were taken simultaneously at measuring points at the 10 kV, 400 V and 690 V levels. As short test sequences during these long-term measurements, particular components and system parts of the power station service were briefly (less than two minutes) switched off and on again. These switching measures enable the effect of certain parts of the system and components to be individually assessed.

Measurement results/assessment The problem in this case can be explained by a single illustration as a documentation of the measurement results. The measurement results of a test sequence are summarised in Figure 7.13.
  This illustration shows the harmonics mean values for the investigated orders in the 5th to 47th order range. The total harmonic distortion is also recorded in this illustration. It can be seen that in normal mode (normal, uncoupled block mode) that the THD of the voltage is 8%. The illustration also shows that the harmonic levels rise significantly towards the higher orders (see 35th and 47th order). When the equipment 'portal building' (intake valve of the power station)

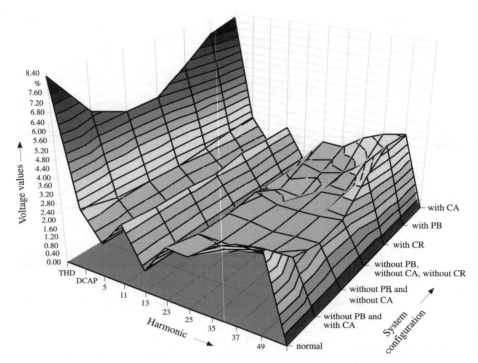

Figure 7.13　*Measuring results of the test sequence on the 400 V voltage level for different operating conditions: PB – Portal Building, CA – Coal Addition, CR – Current Rectifier*

is shut down, there is a very marked reduction in the harmonic voltages in the higher order range and thus also a distinct reduction in the degree of distortion.

Summary/conclusion　In this power station supply network there is a resonant point in the area above the 50th order. The power supply system (the 10 kV incoming cable) of the 'portal building' exercises the essential influence here.

To sum up, it can be said that the aforementioned disturbances are associated with the harmonics distortion of the station service network. The higher order harmonics in particular have a strong disturbing effect on the capacitors used in the network equipment.

If the measuring results are considered from the point of view of standardisation, it can be stated that an assessment of the harmonics up to the 40th order shows that compatibility levels are either just reached or slightly exceeded. From this aspect it is clear that the interference immunity of the relevant equipment is not high enough.

Because of the very short electrical length of the networks in the power station supply networks, the network resonant points generally occur at very high frequencies compared to the medium voltage networks of the public power supply system. While resonant points at harmonic orders in the 7 to 11 range can be expected in the public medium voltage network, the resonant points in the power station supply network are entirely in the 50th to 60th order range, or above.

In the case dealt with here, countermeasures are very difficult to find. The possible measures at the 10 kV voltage level are very cost-intensive. Provided other equipment and devices at other voltage levels do not exhibit failures, the equipment connected to the low voltage supply can be protected by uninterrupt-ible power supplies and a disturbance-free operation thus ensured. It must, however, be considered that the uninterruptible power supplies are operated quasi-permanently on the supply side by a power supply with corresponding system perturbations and therefore must themselves be able to withstand the relevant disturbances.

7.4.2 Disturbance analysis (voltage increase) in power station service network II

Task

The problem of the effects which arise if the interference immunity of equip-ment is not tuned to the disturbance level in the network is clear from the following example. The main problem was seen to be increased difficulties in operating frequency converters which are used with drives of all kinds. The principle cause of disturbances was characterised by an excessive intermediate circuit voltage level. The affected voltage levels are in the 400 V systems.

Data procurement Figure 7.14 shows the system diagram with technical data.

Aspects of measurement To investigate the effect, the harmonics and flicker were measured over a long time period and supplemented by recordings of the time signal. The measurements at the 10 kV, 400 V and 690 V voltage levels were carried out simultaneously.

Measurement results/assessment The measurements showed harmonics levels (2nd to 40th order) in the area of the compatibility levels, as stipulated in the relevant standards.

By comparing the time courses of the voltages on the 10 kV voltage level (see Figure 7.15) with the recordings taken at the most heavily-affected 400 V distri-bution Figure 7.16, different impressions of the commutation occurrences were clearly detected. These are caused by the frequency converters of the boiler-feeding pump drives.

The commutation notches on the 10 kV voltage level show the actual notch which is superimposed by a commutation oscillation ($f = 4$ kHz).

The commutation oscillation is strongly attenuated. The voltage values for the

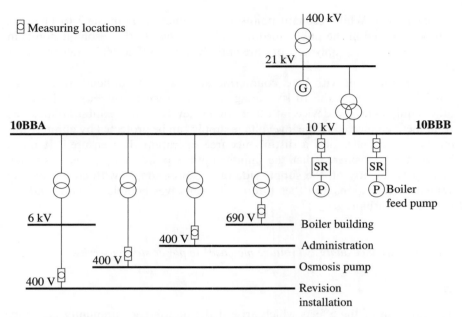

Figure 7.14   *Single-line diagram of medium and low voltage system in a power station*

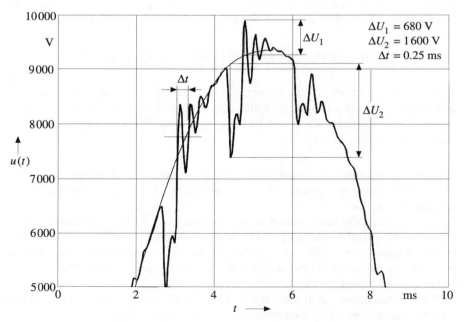

Figure 7.15   *Commutation notches at the 10 kV busbar*

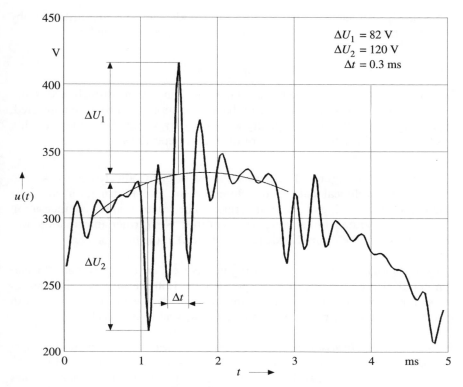

Figure 7.16   Commutation notches at the 400 V busbar

voltage sag and the voltage rise which then occur are within the limits defined by EN 50178 (VDE 0160). An examination of the voltage course at the 400 V busbar reveals some conspicuous features. On the one hand, the commutation notch cannot be very clearly separated from the commutation oscillation and, on the other hand, distinctly higher relative voltage amplitudes occur here. The attenuation of the oscillation is, however, not so pronounced as on the 10 kV side. The deciding factor is that the frequencies of the oscillations are different. In contrast to the 4 kHz frequency at the 10 kV side, the frequency of the commutation oscillation at the 400 V side is approximately 3.3 kHz. This is a clear indication that a resonant point, located at a frequency of 3.3 kHz, is excited by the commutation notch. The oscillation which can be seen at the 400 V side is not connected with the commutation oscillation at the 10 kV side.

Summary/conclusion In the actual case, a resonance excitation is present in a 6 kV level. The converter disturbance signals, which indicate an excessive intermediate circuit voltage level, are caused by the slow charging of the intermediate circuit due to the voltage peaks superimposed on the actual network voltage.

A remedy can be provided in this case, under certain circumstances, by relocating the resonant point in conjunction with an attenuating load.

7.4.3 Network resonance in the low voltage network

Task In an operation disturbed by flicker, in which increasing disturbance excitations and recordings of an uninterruptible power supply (UPS) were registered, an investigation of the voltage quality was to be carried out. The voltage quality was to be quantified and the causes of the flicker determined.

Data procurement Figure 7.17 shows the network or the affected network area. The essential technical data is therefore present. No information could be obtained on the loading of the transformers. The reactive power compensation systems are operated under automatic control. No further information was obtained on the 10 kV supply side (urban station) in the first stage.

Aspects of measurement The measuring points were arranged so that information could also be obtained on the power flow. Long-term measuring instruments for harmonics, flicker and transient recording were installed at both measuring points. The measuring time period was set to one week.

Measurement results/assessment The harmonics measurement in the administration area resulted in a 1.5% to 4.7% voltage distortion (see Figure 7.18).

The flicker measurement (Figure 7.19) showed distinctly higher flicker values at certain time points, but occurring only for relatively short periods. A comparison of the recording with the log of the uninterruptible power supply showed clear agreements. The log of the uninterruptible power supply contains

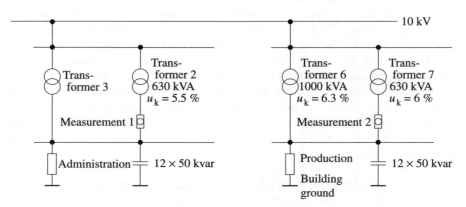

Figure 7.17 Single-line diagram with connection points for measurement in an industrial installation

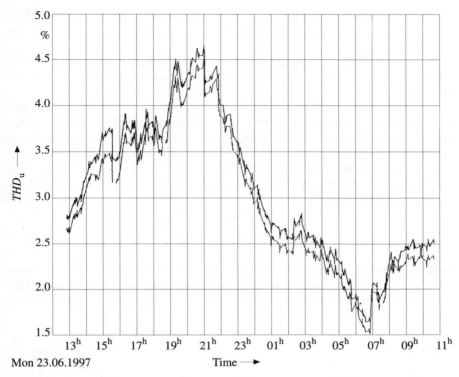

Figure 7.18 Total Harmonic Distortion (THD) of voltage phase-to-earth (U_{L1} and U_{L2})

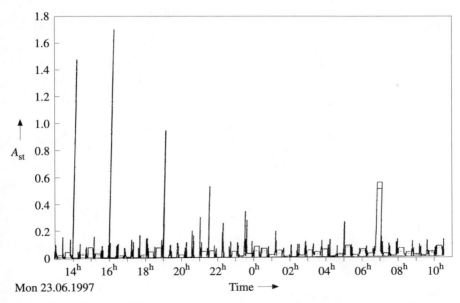

Figure 7.19 Results of flicker measurement of voltage phase-to-earth (U_{L1} and U_{L2})

substantially more recordings but these could not be differentiated further by detail because the recording criteria of the uninterruptible power supply are not transparent.

The transient recordings activated by the voltage trigger correspond in timing with the flicker recordings. Figure 7.20 shows an example taken from a transient recording.

A precise analysis of the transient recording shows that the signals superimposed on the voltage and current have a frequency of 383 Hz. This value corresponds to the telecontrol frequencies of 383.3 Hz in the 10 kV voltage level.

If the system structure in Figure 7.21 is considered, the series resonant circuit which is established, consisting of the transformer inductivity and the capacitance of the reactive power compensation system, can be calculated as follows (see also section 2.3.3):

$$f_{res} = 50 \text{ Hz} \sqrt{\frac{S_r}{u_k} \frac{1}{Q_C}} \tag{7.1}$$

Only six steps of the capacitor bank were investigated, because a closer examination of the reactive power compensation system revealed that six steps had already failed and therefore could no longer be brought on line by the automatic

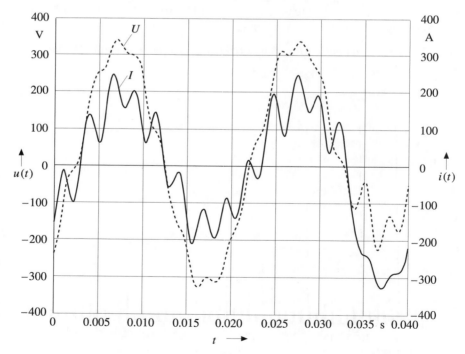

Figure 7.20   Transient recording

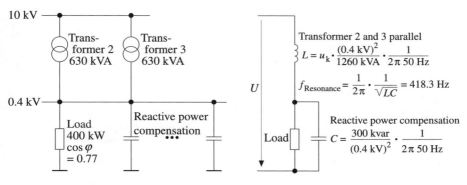

Figure 7.21 Equivalent diagram of series resonant circuit

controller. In this situation, the six steps of the reactive power compensation system still available resulted in a resonant frequency of 418.3 Hz.

Summary/conclusion The telecontrol frequency led to excitation of the series resonant point in the affected low voltage network. The telecontrol energy from the 10 kV network was drained via this series resonance, but the drain was still so slight that it had no effect on the telecontrol operation. However, this level was still sufficient to cause the disturbances mentioned in the task outline in the relevant low voltage network. A remedy can also be found here by relocating the resonant point. In this case it is sufficient to reduce the degree of compensation of the system. With the four steps in operation, it was necessary to allow for a power factor which, although poorer, was still acceptable.

7.4.4 Reactive power compensation in a 500 V network

Task A failure of one stage occurred in an unblocked reactive power compensation system (without blocking reactor) of a 500 V network. The measurement was designed to determine whether an unblocked compensation system could be operated at this network node. In parallel with this, the investigation was also to assess whether the system was designed in accordance with the requirements.

Data procurement The compensation system is constructed without blocking reactor. Its rated voltage is 525 V. The system has a total of six steps each of 55 kvar. Figure 7.22 shows the structure of the system.

Aspects of measurement To investigate the system, the voltages at the connection points and the currents in the branches of the capacitor groups were recorded every second. As part of the measurement, the reactive power compensation system was operated from the shutdown state to step 5. Step 6 was not operable during the measuring time.

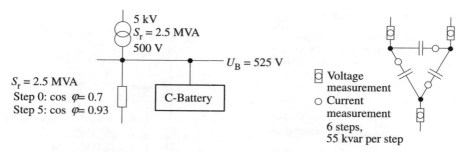

Figure 7.22   Details of single-line diagram at connection point of capacitor bank

Measurement results/assessment The compensation system has a fundamental component rated current in the capacitor branch of approximately 35 A per step. In stage 5, the fundamental component current is about 175 A. The measurement results presented in Figures 7.23 and 7.24 show that the compensation system in the fifth step carries an additional harmonic current of approximately 60 A with the frequency of the 11th harmonic. This current is fed into the network from the existing converters.

The design of the capacitor bank for a rated current of 525 V at normal operating voltage of 525 V means that in its design the capacitor bank is already dimensioned below the expected voltage stress. Considering the existing harmonic currents present in the network, which are drained by the unchoked capacitors, the capacitors are clearly overloaded.

Summary/conclusion Because the capacitor bank is necessary in this system because of the reactive power requirement (see data on cos φ), it is necessary to use a compensation system with blocking reactor. For safety reasons the existing system should no longer be operated. The existing capacitors should also no longer be used because they are certainly already damaged. Furthermore, a distinctly higher voltage rating is necessary for capacitors with blocking reactors.

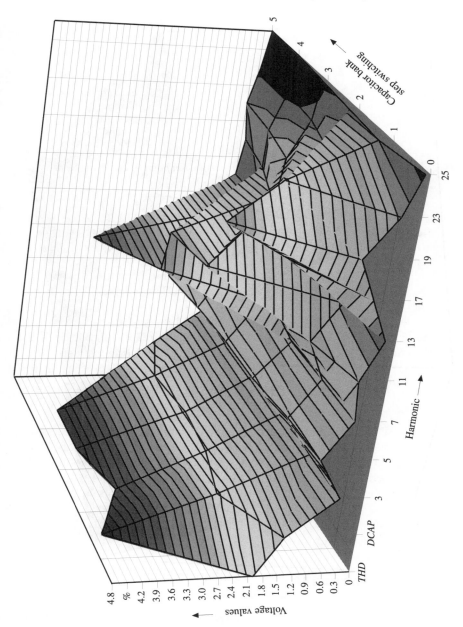

Figure 7.23 Harmonic voltages phase-to-phase for different capacitor steps

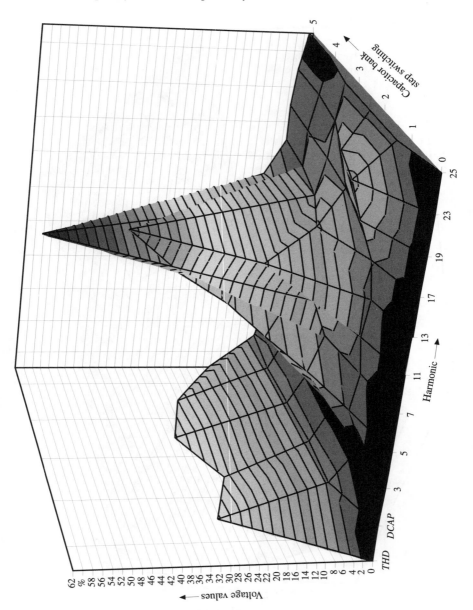

Figure 7.24   Harmonic currents in connection circuits of the capacitor bank for different capacitor steps

7.5 References

1   FGH: 'Oberschwingungsgehalt und Netzimpedanzen elektrischer Nieder-und Mittel- spannungsnetze (Harmonic distortion and network impedances of low and medium voltage networks)'. *Technischer Bericht* No. 1–268, Mannheim 1988

2   SCHLABBACH, J.: 'Netzrückwirkungen bei Anschluß eines Mittelfrequenze-Induktionsofens an ein 10-kV-Netz (System perturbations where a medium-frequency induction furnace is connected to a 10 kV network)'. ETG-Tage 95 (Workshop C), VDE-VERLAG, pp. 221–226

3   BLUME, D., GOEKE, TH., HANTSCHEL, J., PAETZOLD, J., and WELLßOW, W.H.: Einsatzoptimierung von Kondensatorbatterien in einem ausgedehnten 35-kV-Kabelnetz (Application optimisation of capacitor banks in an expanded 35 kV cable network)'. *Elektrizitätswirtschaft* 95, 1996, vol. 8, pp. 474–482

Chapter 8

Appendix

8.1 Formula symbols and indices

8.1.1 Formula symbols

A	Area
A	Flicker disturbance value, annoyance value
a, a²	Rotational phasors
a, b, c	Fourier coefficients
B	Susceptance
B	Magnetic flux density
B	Harmonic distortion factor
B	Bandwidth
C	Capacitance
D	Distortion power
d	Distortion factor
A	Attenuation
d	Voltage change
d	Interest rate
E	Identity matrix
F	Transmission function
F	Form factor
G	Conductance
g	Fundamental component content
H	Magnetic field strength
h	Harmonics order
I	Current, general
J	Current density
k	Degree of asymmetry
k	Factor
k	harmonic content
L	Inductivity

l	Factor
M	Mechanical torque
m	Factor
N	Number of turns, factor
n	Speed of rotation
P	Active power
P	Disturbance value
$PWHD$	Partial Weighted Harmonic Distortion
p	Number of pulses
Q	Reactive power
q	Number of windings
R	Resistance
r	Repetition rate
r	Reduction factor
S	Apparent power
s	Length
T	Time, time instant
THD	Total Harmonic Distortion
TIF	Telephone Interference Factor
T	Transformation matrix
t	Time duration, time course
U	Voltage, general
$ü$	Angle of overlap
V	Costs, value
X	Reactance
Y	Admittance
Z	Impedance
α	Control angle
δ	Loss angle
Θ	Magnetomotive force
Θ	Moment of inertia
ϑ	Pole angle
ϑ	Temperature
ψ	Impedance angle
λ	Power factor
μ	Factor
τ	Time constant
Φ	Magnetic flux
Φ	Luminous flux
φ	Angle, load angle
ω	Angular velocity

8.1.2 *Indices, subscript*

A	Connection
A, B, C	General index
B	Reference value
b	Reactive component
C	Capacitor, capacitive
c	Critical
D	Reactance coil
D, d	Delta winding
d	Direct axis
d	Direct voltage
d	Dielectric
E	Earth
f	Function
fe	Weighted factor
G	Generator
Neg	Negative sequence system quantity
Tot	Total
HV	High voltage
h	Harmonics order
I	Current
i	Part
j	Part
k	Degree of asymmetry, short-circuit
k3	Three-phase short-circuit
L	Inductivity, inductive
L	Lamp
L	Line
L	Load-side
$L_1/L_2/L_3$	Three-phase components
LV	Low voltage.
lt	Long-term value
M	Motor
MV	Medium voltage
N	Supply-side
LV	Low voltage
max	Maximum value
Pos	Positive sequence system quantity
n	Nominal value
OV	Higher voltage side
p	Pole spider
ph	Phase position
Q	Supply point
R, Y, B	Three-phase components

R	Ohmic
Rest	Residual
r	Rated value
res	Resonance
s	Secondary
s	Transmitter
St	Converter
st	Short-term value
S	Switching frequency
T	Transformer
t	Time instant
U	Voltage
UV	Lower voltage side
V	Losses
V	PCC
VT	Compatibility
v	Prohibited
W	Coil
w	Active component
Y, y	Star winding
Int	Interharmonic
m	Magnetisation
s	Spread
1	Fundamental frequency
0, 1, 2	Symmetrical components
$+, -$	Limit frequency

8.1.3 Indices, superscript

$''$	Subtransient
$*$	Conjugated complex
$'$	Relative, p.u.

Identification, U as example

$\underline{U}$	Complex quantity
U	Effective value of a sinusoidal, time-dependent quantity
$|\underline{U}|$	Amount of a complex quantity
$\mathbf{U}$	Matrix, vector
u	Instantaneous value, quantity which changes over time, relative quantity
$\underline{U}^*$	Conjugated complex quantity
$u(t)$	Quantity which changes over time
U_h	Effective value of the magnitude of the harmonics order h

Sequence of subscripted indices

First position:	Component	U_R or U_1
next position:	Operating state	U_{Rk}
or:	Harmonics order	U_{Rh}
next position:	Type of equipment	U_{RkT}
next position:	Number of equipment	U_{RkT3}
next position:	Additional designation	$U_{RkT3max}$
next position:	Running index	$U_{RkT3maxi}$

Index